STATICS WORKBOOK

Making Statics Permanent for the Emerging Engineer

AF892218

Kendall Hunt
publishing company

Jamie
Gurganus

Cover image: © Shutterstock, Inc.

Kendall Hunt
publishing company

www.kendallhunt.com
Send all inquiries to:
4050 Westmark Drive
Dubuque, IA 52004-1840

Copyright © 2016 by Kendall Hunt Publishing Company

ISBN 978-1-4652-9440-1

All rights reserved. No part of this publication may be reproduced, stored in a retrieval system, or transmitted, in any form or by any means, electronic, mechanical, photocopying, recording, or otherwise, without the prior written permission of the copyright owner.

Printed in the United States of America

CONTENTS

Acknowledgments

Richard Blorstad – Essential Contributing Author and Illustrator

Problem Authors:

Jorden Bendler
Derek S. Bibeault
Kurtis Boulter
Megan Centafont
Keith Coleman
Niara Comrie
Nicole Demetrides
Brett Geraghty
Sarah Hemler
Garrett Higgins
Robbie Oehrli
Dr. Neil Routhman
Jordyn Schroeder
Jonathan Shulgach
Tess Wootton-Klabanoff
Caroline Zelhofer
Yangling Zhou

Chapter 1

Static Math

Yep, that stuff you did in high school matters

INTRODUCTION

Think back to your high school math classes. Remember when you were wondering when any of this even mattered? Well, here's your answer! Engineering, huzzah!

In your engineering courses, math is an important tool in obtaining correct solutions for the problems. Statics is no exception. The student must have a solid base in several mathematical disciplines to set up and solve the equations of statics.

Algebra, trigonometry, geometry, and calculus are all very important in the study of statics.

In this section, it is intended that the student will *review* the basic mathematical principles that are used extensively in this course.

Before beginning the course, it is *highly encouraged and recommended* that you review and master your math skills before diving into engineering statics.

Before we get started, it is important to understand in statics, we use a *right-handed coordinate system*. A perfect example of such a coordinate system is the Cartesian coordinate system, as shown.

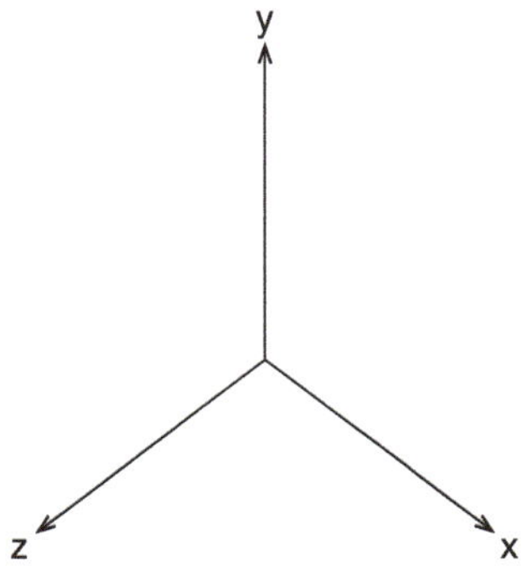

TRIGONOMETRY NEEDED FOR STATICS

Basic trigonometric functions: The three basic trig functions used in statics are sine, cosine, and tangent.

Consider the triangle's in the following:

For a *right triangle* (with a 90° angle):

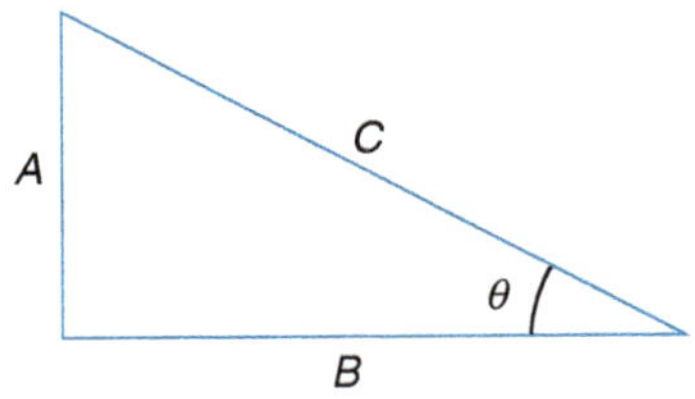

Pythagorean Theorem: $C^2 = A^2 + B^2$

$$\sin(\theta) = \frac{A}{C} = \frac{Opposite}{Hypotenuse}$$

$$\cos(\theta) = \frac{B}{C} = \frac{Adjacent}{Hypotenuse}$$

$$\tan(\theta) = \frac{A}{B} = \frac{Opposite}{Adjacent}$$

For any *general triangle* (including right triangles):

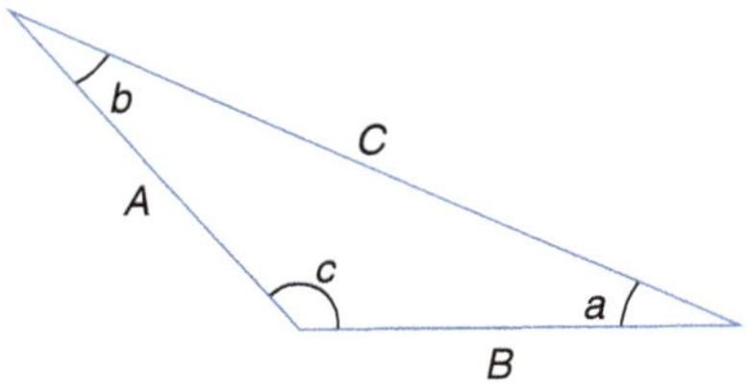

Law of Sines

$$\frac{A}{\sin(a)} = \frac{B}{\sin(b)} = \frac{C}{\sin(c)}$$

Law of cosines:

$$C^2 = A^2 + B^2 - 2AB\cos(c)$$

Recall that for any triangle, the sum of the angles always is 180°

$$a + b + c = 180°$$

GEOMETRY

There are several geometric rules/theorems that you will utilize in statics to solve problems. If you ever wondered why you had to memorize some of these theorems, your answer may lie in the following chapters, but first we will have to go over them again.

Opposite Angle Theorem (Vertically Opposite Angles)

Through any two lines that intersect, the angles that are opposite one another are equal.

$$\theta = \theta$$

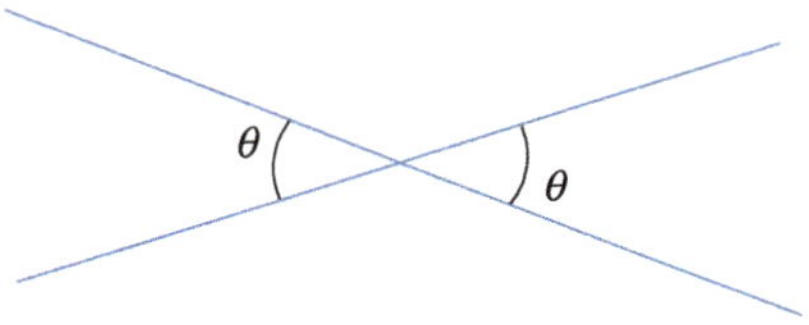

Definition of Supplementary Angles

Any two angles that have a side on the same line are supplementary.

$$\theta + \phi = 180^\circ$$

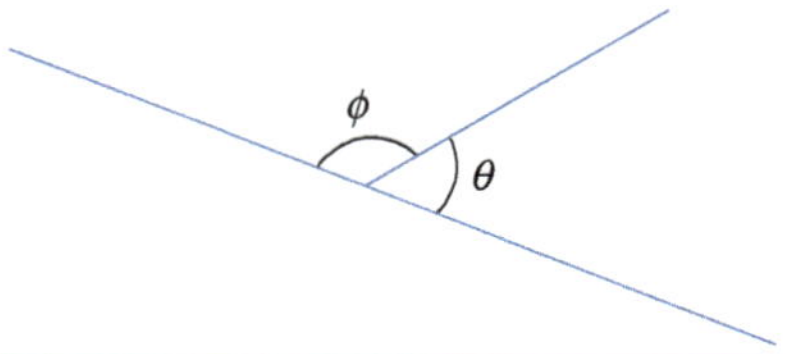

Similar Triangles

In similar triangles, the respective angles are equivalent, despite the fact that the sides are different. Similar triangles are defined where,

$$\frac{A_1}{A_2} = \frac{B_1}{B_2} = \frac{C_1}{C_2}$$

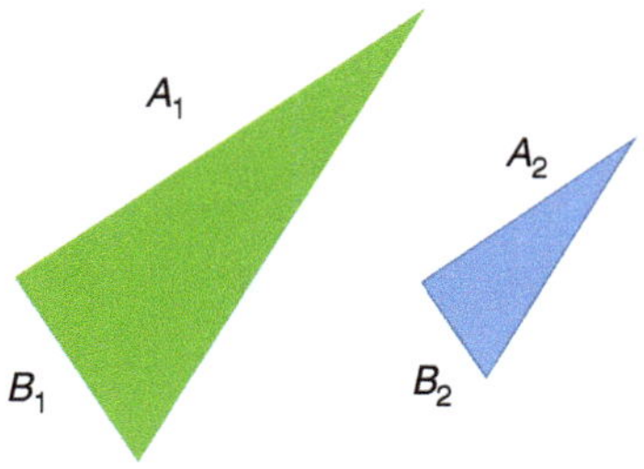

Properties of Common Shapes

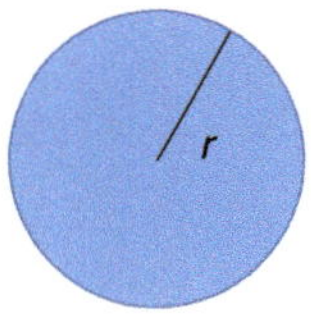

Area $A = \pi r^2$
Circumference $C = 2\pi r$

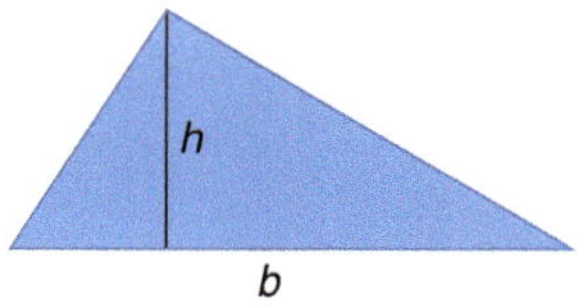

Area $A = \frac{1}{2}bh$

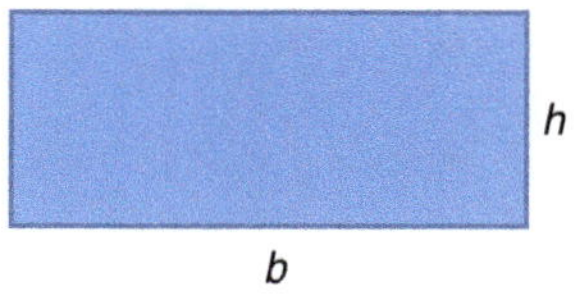

Area $A = bh$

Note: For some "Tricks" in Trig, see the Appendix A.

SECTION 1 Problems

1.1. BASIC TRIGONOMETRY

Problem Author: Brett Geraghty

Type of Problem:
This is a review sheet to go over basic trigonometric identities and useful relationships.

BASIC TRIG IDENTITIES

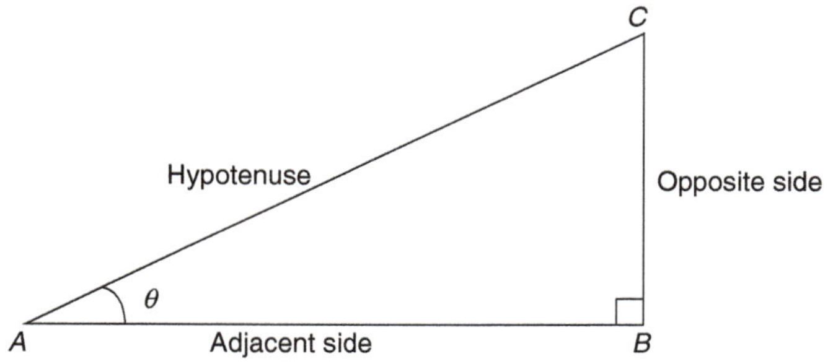

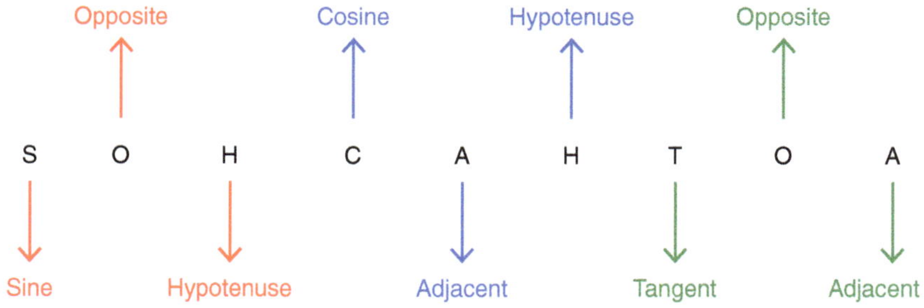

$$\sin(\theta) = \frac{\textit{Opposite side}}{\textit{Hypotenuse}} \qquad \cos(\theta) = \frac{\textit{Adjacent side}}{\textit{Hypotenuse}} \qquad \tan(\theta) = \frac{\textit{Opposite side}}{\textit{Adjacent side}}$$

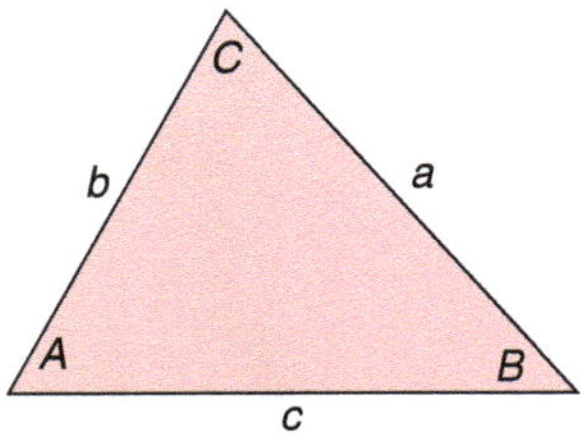

Law of Sines

$$\frac{a}{\sin(A)} = \frac{b}{\sin(B)} = \frac{c}{\sin(C)}$$

Law of Cosines

$$a^2 = b^2 + c^2 - 2bc \times cos(A)$$

$$b^2 = a^2 + c^2 - 2ac \times cos(B)$$

$$c^2 = a^2 + b^2 - 2ab \times cos(C)$$

Practice and Application:

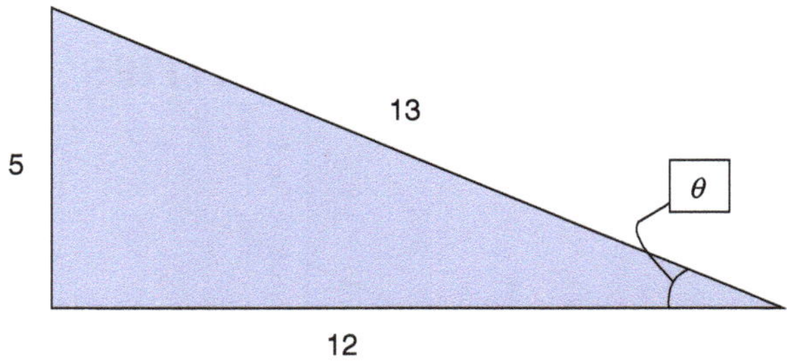

1. Find: $\sin(\theta)$, $\cos(\theta)$, $\tan(\theta)$, and θ

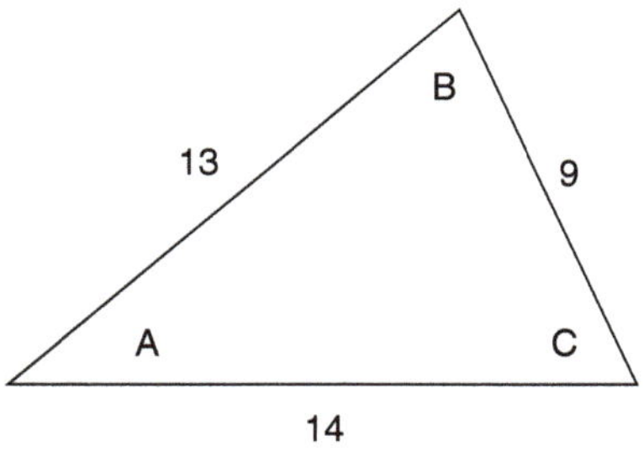

2. Use the law of cosines to determine angles A, B, and C

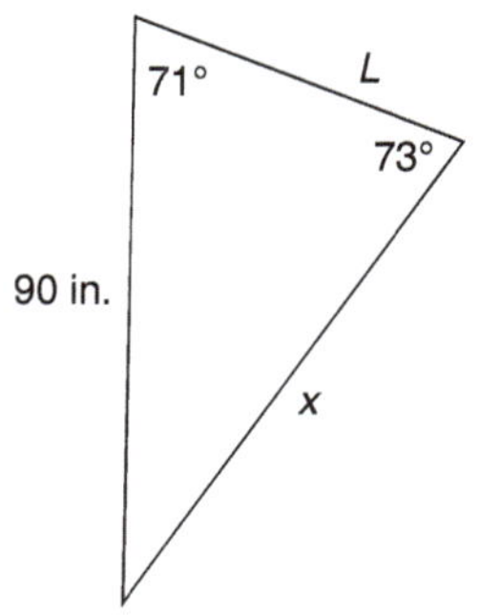

3. Use the law of sines to determine x and L

Solutions:

1)

$$\sin(\theta)=\frac{5}{13}\approx 0.385 \quad \cos(\theta)=\frac{12}{13}\approx 0.923 \quad \tan(\theta)=\frac{5}{12}\approx 0.417 \quad \theta\approx 22.620^\circ$$

2) $a = 9$, $b = 14$, $c = 13$

$81=196+169-(2\times 13\times 14)\cos(A)$ $\quad \cos(A)=\frac{71}{91}$ $\quad A\approx 38.719^\circ$

$196=81+169-(2\times 9\times 13)\cos(B)$ $\quad \cos(B)=\frac{3}{13}$ $\quad B\approx 76.658^\circ$

$169=81+196-(2\times 9\times 14)\cos(C)$ $\quad \cos(C)=\frac{3}{7}$ $\quad C\approx 64.623^\circ$

As a check, the sum of the angles should be equal to 180° $\quad A+B+C=180^\circ$

3)

$$\frac{90}{\sin(73^\circ)}=\frac{x}{\sin(71^\circ)} \qquad x\approx 88.985 \text{ in.}$$

The third angle is $180^\circ-71^\circ-73^\circ=36^\circ$ $\quad \frac{90}{\sin(73^\circ)}=\frac{L}{\sin(36^\circ)}$ $\quad L\approx 55.318$ in.

1.2. TRIGONOMETRY OF AN ISOSCELES TRIANGLE

Problem Author: Richard Blorstad

Type of Problem:
The student is asked to identify trigonometric/geometric properties to determine the height of the triangle.

Problem Statement:
Given the following figure, solve for the height, h, of the triangle using what you know about geometry and trigonometry.

Drawing:

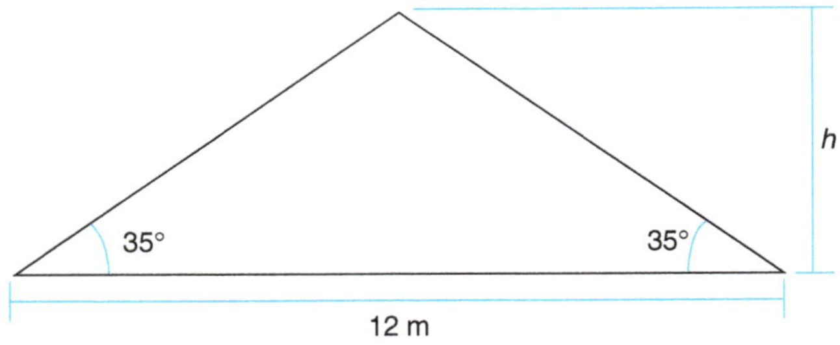

Solution:
Since it is an isosceles triangle, the height can be determined using a single trigonometric identity.

$$h = \frac{12}{2} \times \tan(35°)$$

$$h = 4.2 \text{ m}$$

Goal:
To be able to break the isosceles triangle into a right triangle, thus simplifying the problem.

Given:
Angle (θ) = 35°, Length = 12 m

Assumptions:
None

Equations Needed:

$$\tan(\theta) = \frac{Opposite}{Adjacent}$$

1.3. TRIGONOMETRY IN BASEBALL

Problem Author: Garrett D. Higgins

Type of Problem:

These problems cover trigonometry and its applications in statics (and all other branches on engineering). Students will learn how to apply the law of sines, the law of sines, and any other type of trigonometric functions and identities as well as the parallelogram law and of important geometry principals. By the end of this section, students should be able to recognize where trigonometry can be utilized, recall identities and formulas relating to trigonometric functions, and apply them to solve problems.

Problem Statement:

Question 1: During a baseball game, the batter hits the ball at an angle of 37° to the horizon. The ball then flies to a maximum height of 90 ft. above the ground before falling back to Earth; what is the x-component of the distance from the batter to the ball, when the ball is at maximum height? Assume the ball travels linearly after contacting the bat.

Drawing:

If possible, please take an original photograph or draw as clear as possible the problem. Include any necessary constraints.

Solution:

119.43 ft.

Goal:

The goal here is to recognize that at the maximum height of the ball, a right angle is formed between the batter and the ball.

Given:

$\Theta = 37°$, Y-component of height is 90 ft.

Assumptions:

The answer can be found using the tangent function.

Equations Needed:

$\tan \Theta = (Opposite/Adjacent)$

Answer:

$$\tan\Theta = \left(\frac{o}{a}\right)$$

$$\tan\Theta = \frac{90\text{ ft.}}{x}$$

$$x = \frac{90\text{ ft.}}{\tan(37°)}$$

$$x = 119.43\text{ ft.}$$

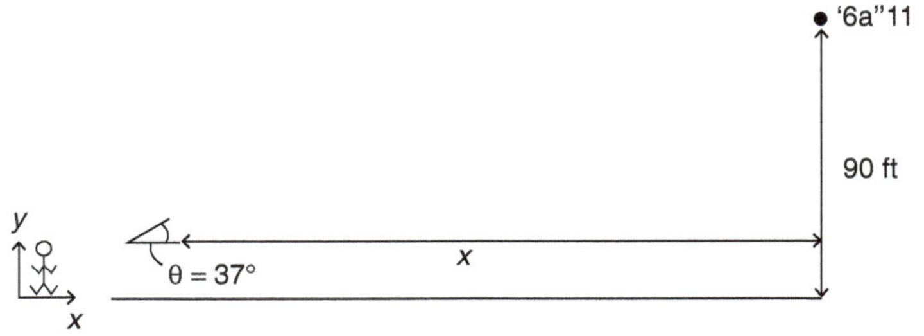

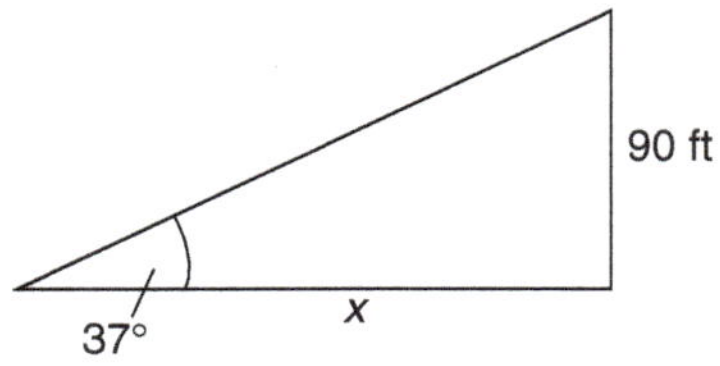

$$\tan\Theta = \frac{\text{Opposite}}{\text{Adjacent}}$$

$$\tan\Theta = \frac{90\text{ ft}}{x}$$

$$x\tan\Theta = 90\text{ ft}$$

$$x = \frac{90\text{ ft}}{\tan\Theta}$$

$$x = \frac{90\text{ ft}}{\tan 37°}$$

$$x = 119.43\text{ ft}$$

1.4. TRIGONOMETRY OF MULTIPLE TRIANGLES

Problem Author: Richard Blorstad

Type of Problem:
Math in statics, trigonometry

Problem Statement:
An English ship is being chased by pirates. You are standing atop the cliff, watching the events unfold beneath you. Using some basic geometric principles, you determined the angles in the following picture and using nothing more than a rock, a digital stopwatch (because those definitely existed in the eighteenth century), and your knowledge of gravity and physics, you determined the height of the cliff to be approximately 100 ft. Determine the distance between the ships.

Drawing:

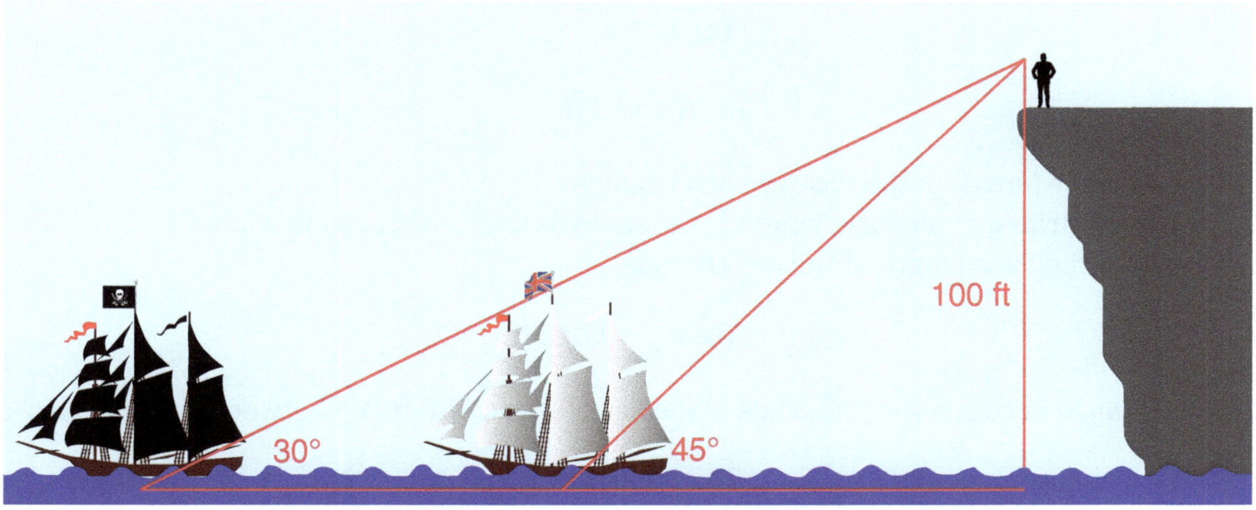

Courtesy of Richard Blorstad

Goal:
Find the distance between the two ships using trigonometry.

Given:
The two right triangles and the common height of the triangles

Assumptions:
None

Equations Needed:

$$\frac{\sin(A)}{a} = \frac{\sin(B)}{b} = \frac{\sin(C)}{c}$$

$$c^2 = a^2 + b^2,\ \sin(\theta) = \frac{Opposite}{Hypotenuse},\ \cos(\theta) = \frac{Adjacent}{Hypotenuse},\ \tan(\theta) = \frac{Opposite}{Adjacent}$$

Answer:

Can be solved (at least) two ways

1) This problem requires a few steps. First, find the hypotenuse of the small right triangle.

$$\text{Hypotenuse}_{45^\circ} = 100 \times \sin(45^\circ) = 141.4\ \text{ft.}$$

Second, determine the angle between the two diagonal lines (the small angle in the top right of the obtuse triangle). *Note:* the two acute angles that make up a right triangle must be complimentary.

$$60^\circ - 45^\circ = 15^\circ$$

Last, use law of sines to determine the distance between the ships (d).

$$\frac{\sin(30^\circ)}{141.4} = \frac{\sin(15^\circ)}{d}$$

$$d = 73.2\ \text{ft.}$$

2) First, find the length of the base of each triangle.
For the 45° triangle, the base length is the same as the height, 100 ft.
Next, find the base length of the 30° triangle.

$$\text{Base}_{30^\circ} = 173.2\ \text{ft.}$$

The distance between the two boats is the same as the difference between the triangles

$$173.2 - 100 = 73.2\ \text{ft.}$$

1.5. TRIGONOMETRY IN THE REAL WORLD

Problem Author: Sarah Hemler

Type of Problem:
Students should be able to use law of sines or cosines and geometry to find specific angles.

Problem Statement:
Find angles $\overline{DBC}$, $\overline{BCD}$, and $\overline{BAE}$ in the following symmetrical swing set frame by using the geometry.

Drawing:

Courtesy of Sarah Hemler

Solution:

$\overline{DBC} = 28.96°$ $\overline{BCD} = 46.58°$ $\overline{BAE} = 64.44°$

Goal:
Find angles $\overline{DBC}$, $\overline{BCD}$, and $\overline{BAE}$

Given:

Geometry

Assumptions:

Frame is stationary.

Equations Needed:

$$\frac{\sin a}{A} = \frac{\sin b}{B} \qquad C^2 = A^2 + B^2 - 2AB\cos c$$

Answer:

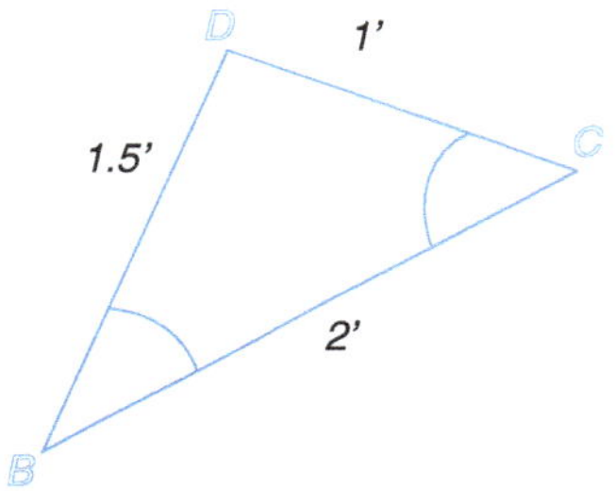

Using law of cosines, find angle $\overline{DBC}$:

$$|DC|^2 = |DB|^2 + |BC|^2 - 2|DB||BC|\cos\overline{DBC}$$

$$\overline{DBC} = \cos^{-1}\left(\frac{|DB|^2 + |BC|^2 - |DC|^2}{2|DB||BC|}\right) = \cos^{-1}\left(\frac{1.5\text{ ft.}^2 + 2\text{ ft.}^2 - 1\text{ ft.}^2}{2\times 1.5\text{ ft.}\times 2\text{ ft.}}\right) = 28.96°$$

Using law of sines, find angle $\overline{BCD}$:

$$\frac{\sin\overline{DBC}}{|DC|} = \frac{\sin\overline{BCD}}{|BD|}$$

$$\overline{BCD} = \sin^{-1}\left(|BD|\times\frac{\sin\overline{DBC}}{|DC|}\right) = \sin^{-1}\left(1.5\text{ ft.}\times\frac{\sin 28.96°}{1\text{ ft.}}\right) = 46.58°$$

Find angle $\overline{BAE}$:

First, find angle $\overline{BDC}$:

$$180° - 28.96° - 46.58° = 104.46°$$

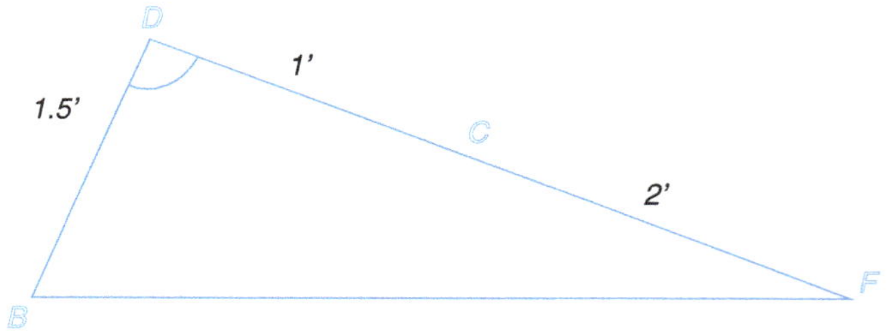

Use law of cosines to find $|BF|$

$$|BF|^2 = |BD|^2 + |DF|^2 - 2|BD||DF|\cos\overline{BDC}$$

$$|BF| = \sqrt{1.5\text{ ft.}^2 + 3\text{ ft.}^2 - 2\times 1.5\text{ ft.}\times 3\text{ ft.}\cos(104.46°)} = 3.22\text{ ft.}$$

Use law of sines to $\overline{DBF}$

$$\frac{\sin\overline{BDC}}{|BF|} = \frac{\sin\overline{DBF}}{|DF|}$$

$$\overline{DBF} = \sin^{-1}\left(|DF|\times\frac{\sin\overline{BDC}}{|BF|}\right) = \sin^{-1}\left(3\text{ ft.}\times\frac{\sin 104.46°}{3.22\text{ ft.}}\right) = 64.44°$$

According to the parallelogram law, angles $\overline{DBF}$ *and* $\overline{BAE}$ *are equal:*

$$\overline{BAE} = 64.44°$$

CHAPTER 2

Forces (2D and 3D Representation)

I can't take the pressure!

INTRODUCTION

So, what do we do with all this math? Let's calculate forces! In this section, you will review:

1. Definition of a force.
2. External versus internal forces (sometimes they are on the outside and sometimes inside).
3. Forces (just another word for a vector) and what's the deal with these unit vectors.
4. What if I want to find more forces?

FORCES

External Versus Internal Forces

In a system, a force can be either external or internal. There are several types of forces that are either external or internal to include the following: gravitational, normal, friction, fluid, tension, compressive, and shear. It is important for an engineering student to be able to classify what forces are on or within a system and whether they are either external or internal.

External forces act on the system whereas internal forces act within the system.

? My name is Force (aka Vector)

Why do we need these "vectors"?

In statics, we study forces. Forces are vectors that have both *magnitude*, size, the strength of the force, and *direction*, which is relative to the coordinate plan. See figure in the following.

Scalars: only contain magnitude and its unit
Units of force

- Newtons (N) in SI units
- Pounds (lb.) in US Customary units

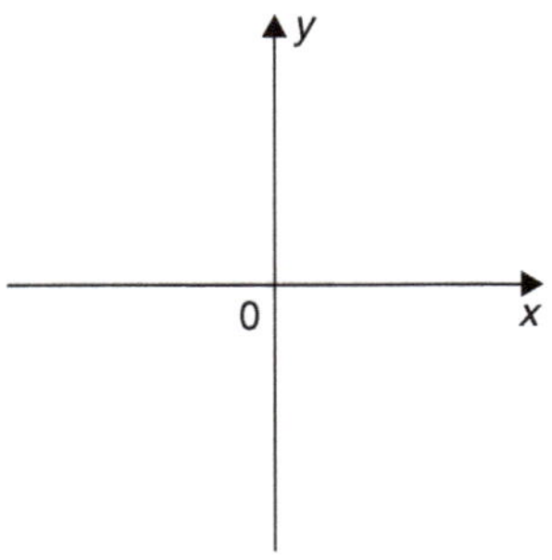

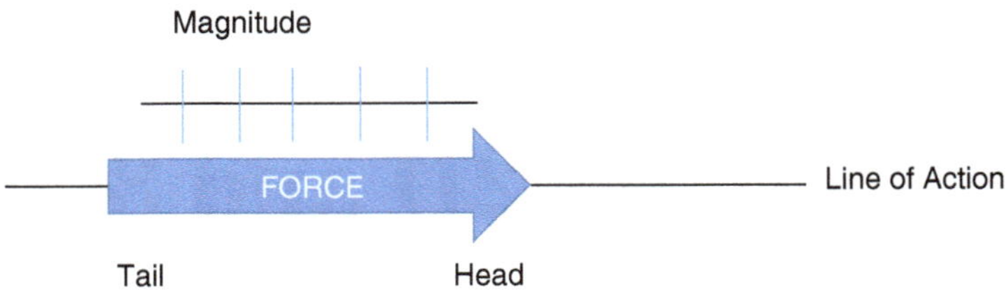

The above figure shows a coordinate plane with force acting to the right in the positive x-direction. You can move this force anywhere along the line of action without changing the effect it has on the system.

What's the deal with Unit vectors

Unit vectors are defined as vectors whose length is exactly one. This means you can use a unit vector to identify the direction and then assign a length (magnitude) to the vector to get a unique vector.

Three Unit Vectors

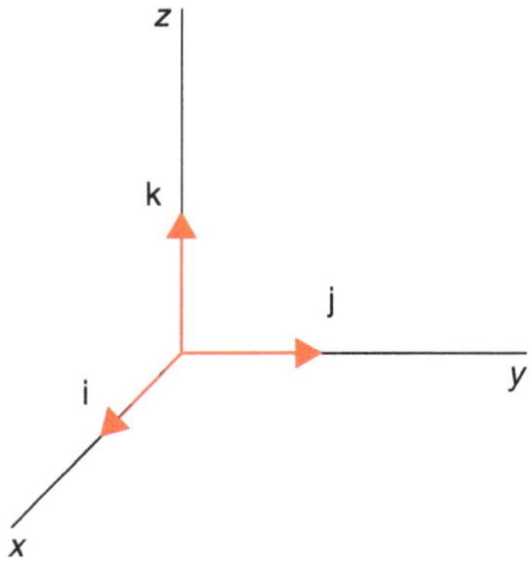

i the unit vector in the direction of the x-axis
j the unit vector in the direction of the y-axis
k the unit vector in the direction of the z-axis

SOLVING FOR A FORCE'S MAGNITUDE AND DIRECTION (2D AND 3D)

How can we find the magnitude and direction of a Force?

To solve for forces magnitude and direction, you will need to be able to use the following mathematical tools:

- Trigonometry (reviewed in Chapter 1)
- Vector addition (parallelogram law)
- Dot product

Each of these will be reviewed in the following.

To find the *magnitude* and *direction* of a force (vector), the following can be used depending on the information supplied in a problem:

A. The *angles* the force creates with each axis in a right-handed coordinate system

$$\left.\begin{aligned}\theta_x &= \cos^{-1}\left(\frac{x_2 - x_1}{L}\right)\\ \theta_y &= \cos^{-1}\left(\frac{y_2 - y_1}{L}\right)\\ \theta_z &= \cos^{-1}\left(\frac{z_2 - z_1}{L}\right)\end{aligned}\right\}$$

These angles are known as **SPACE ANGLES** or **COORDINATE ANGLES**

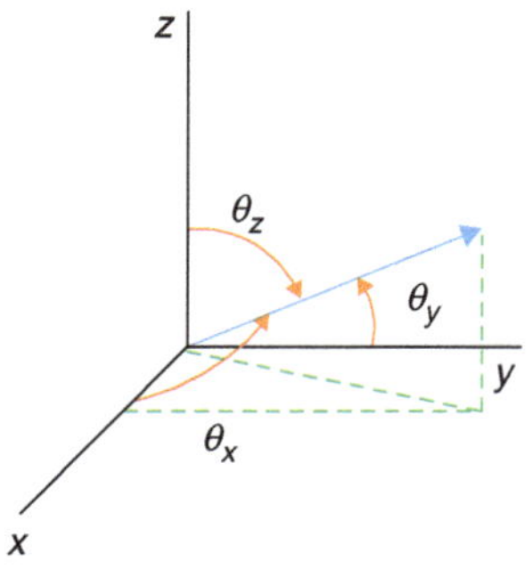

$$L = \sqrt{(x_2 - x_1)^2 + (y_2 - y_1)^2 + (z_2 - z_1)^2}$$

This is the **MAGNITUDE (length)**

Space angles: Are related and NOT independent of one another. They are related by their direction cosines:

$$\sqrt{(\cos\theta_x)^2 + (\cos\theta_y)^2 + (\cos\theta_z)^2} = 1$$

These cosine terms are known as the **DIRECTION COSINES**

B. The *angles* the force forms with a *spherical* coordinate system

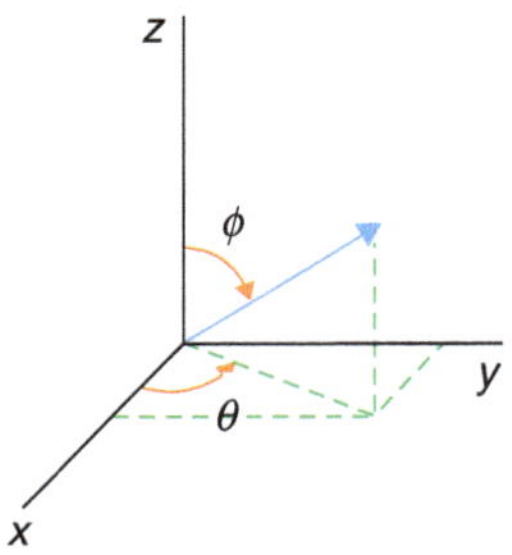

$$\begin{aligned}\cos\theta_x &= (\sin\phi)(\cos\theta)\\ \cos\theta_y &= (\sin\phi)(\sin\theta)\\ \cos\theta_z &= \cos\phi\end{aligned}$$

C. Rectangular component representation: The vector sum of the force components or taking components

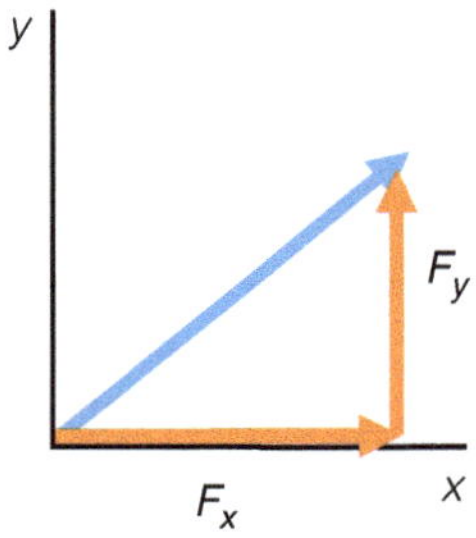

Recall:

2D form Vectors are then written

$$\boldsymbol{F} = \boldsymbol{F}_x\ \boldsymbol{i} + \boldsymbol{F}_y\ \boldsymbol{j}$$

3D Vectors are then written

$$\boldsymbol{F} = \boldsymbol{F}_x\ \boldsymbol{i} + \boldsymbol{F}_y\ \boldsymbol{j} + \boldsymbol{F}_z\ \boldsymbol{k}$$

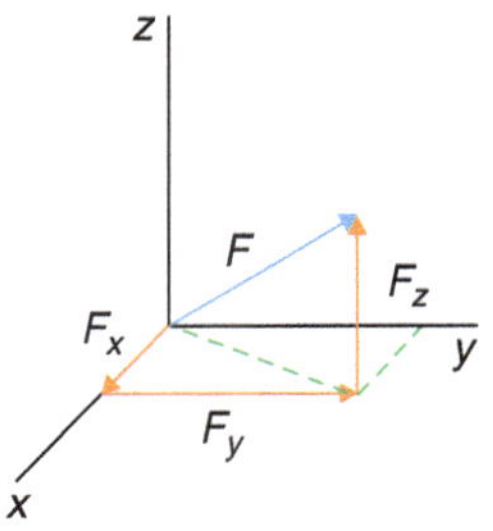

If you know the *magnitude* and *angles*, you can solve for the components in each direction

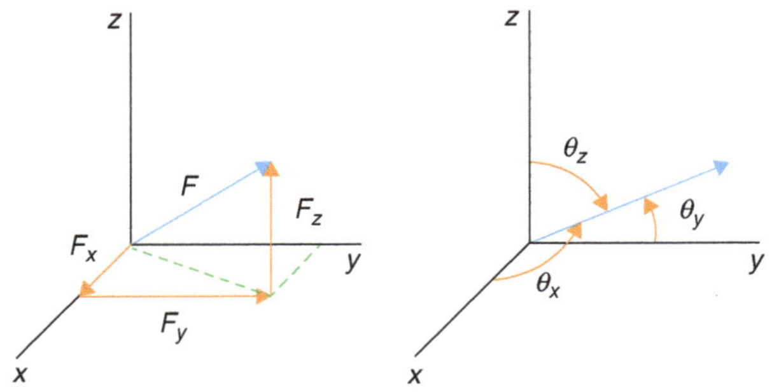

$$F_x = ||F||\cos\theta_x$$

$$F_y = ||F||\cos\theta_y$$

$$F_z = ||F||\cos\theta_z$$

D. The unit vector that acts along the line of action of the force
The following shows graphically how to solve for unit vectors using force components.

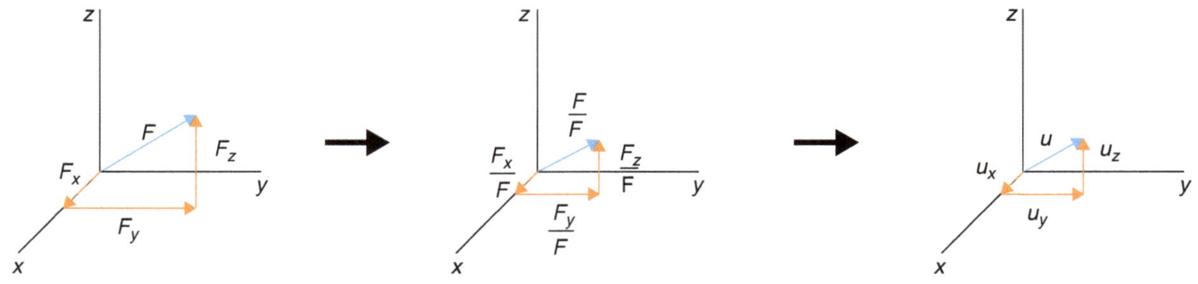

Force Vector and Components

Dividing by the Force Magnitude

Realizing that $\frac{F}{F} = 1$, we now have the unit vector, u

$$u = \frac{F}{||F||} = \frac{||F||\cos\theta_x\ \boldsymbol{i} + ||F||\cos\theta_y\ \boldsymbol{j} + ||F||\cos\theta_z\ \boldsymbol{k}}{||F||}$$

This results in the unit vector, u

$$u = \cos\theta_x\ \boldsymbol{i} + \cos\theta_y\ \boldsymbol{j} + \cos\theta_z\ \boldsymbol{k}$$

Recall from before that $\sqrt{(\cos\theta_x)^2 + (\cos\theta_y)^2 + (\cos\theta_z)^2} = 1$

A second way to solve for the unit vector is using position vector r. This is shown graphically on the following page. Position vector is discussed more in Chapter 3.

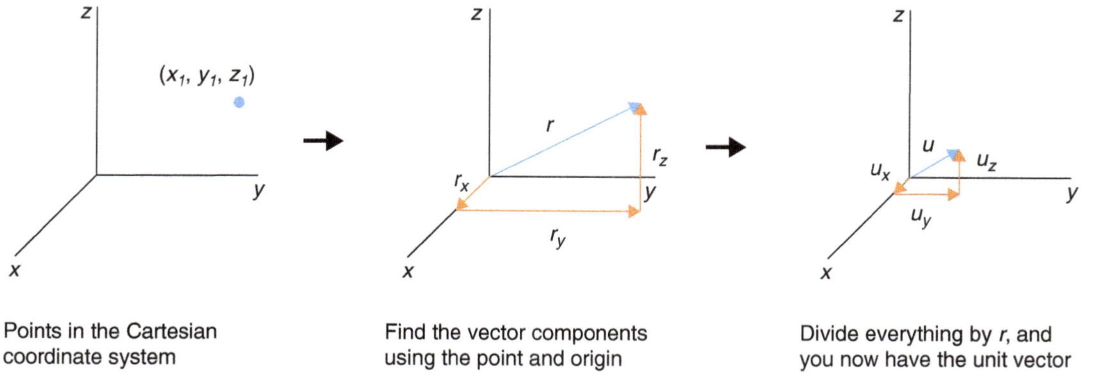

A third way to solve for the unit vector, is using space angles. The following shows this graphically.

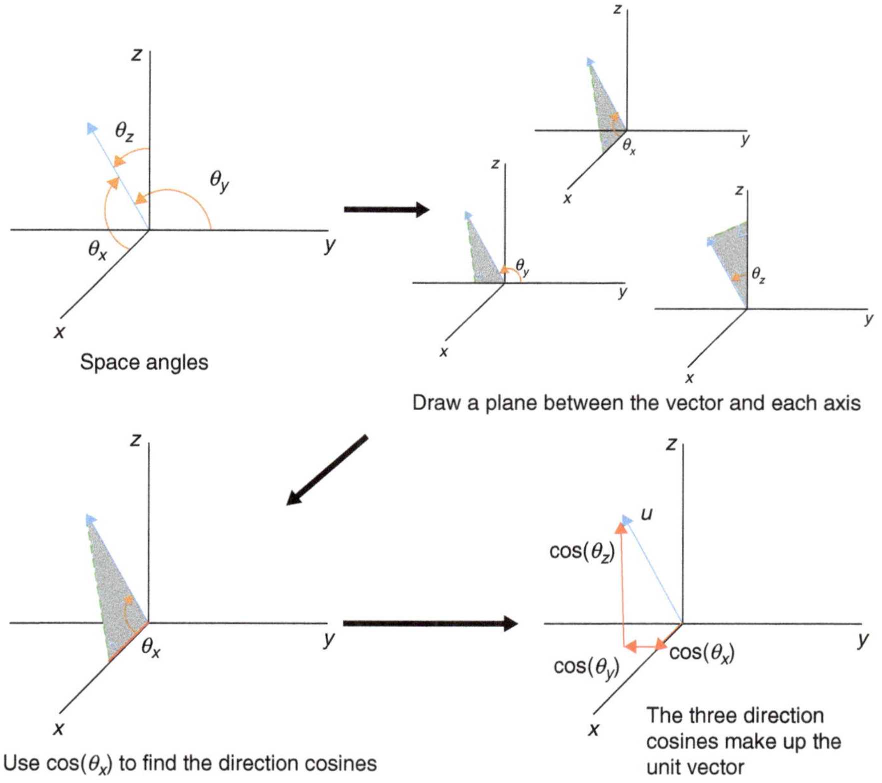

What if I want to find more forces?

Vector Addition—Resultant Force

Force vectors can be added several ways, which include:

- Graphical Addition: head to tail approach, parallelogram law of addition
- Geometric/Trigonometric Addition
- Force (Vector) Component Addition

The first two methods will not be summarized in this book.

Force (Vector) Component Addition Method

- Break each vector down into its components
- Add up the components of n vectors

$$\boldsymbol{F}_R = \Sigma \boldsymbol{F} = (\Sigma \boldsymbol{F}_x)\boldsymbol{i} + (\Sigma \boldsymbol{F}_y)\boldsymbol{j} + (\Sigma \boldsymbol{F}_z)\boldsymbol{k}$$

$$\boldsymbol{F}_{R_x} = \Sigma \boldsymbol{F}_{n_x}$$

- Repeat that for the y and z components of the resultant
- Use the Pythagorean theorem to get the resultant magnitude and trigonometry to get the direction

Dot Product

The Dot product is useful for a number of applications with force vectors to include solving for angles between forces and projection of a force.

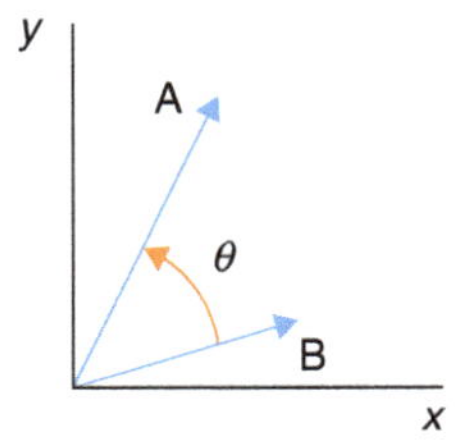

Solving for angles between forces

$$A \cdot B = AB\cos(\theta)$$

$$= A_x B_x + A_y B_y + A_z B_z$$

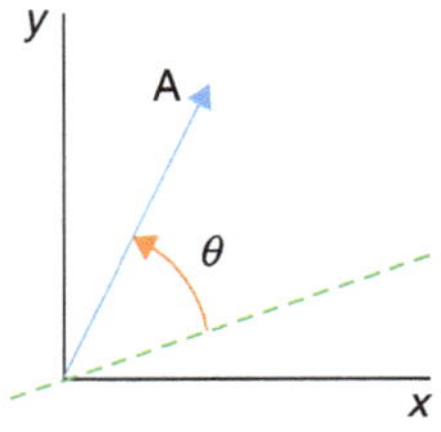

Projection of a force

$$A \cdot u = A \cos(\theta)$$

SECTION 2 Problems

2.1. 3D FORCE REPRESENTATION ON A TENT STAKE

Problem Author: Robert Oehrli

Type of Problem:
3D force components

Problem Statement:
The rope tied around the tent steak has a 30 N tension force. Calculate the components of the force.

Drawing:

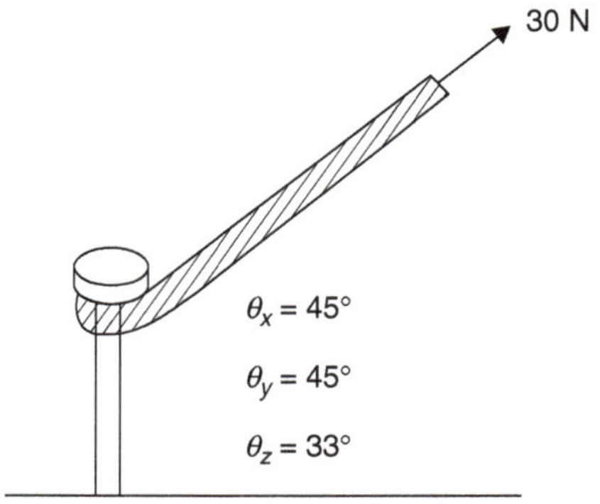

Goal:
Practice breaking down 3D forces vectors into components.

Given:

$$\theta_x = 45°$$

$$\theta_y = 45°$$

$$\theta_z = 33°$$

Solution:

Since the Cartesian angles are given, start by figuring out if the x, y, and z components represent the adjacent side or the opposite side of the triangle. If the component represents the opposite side of the triangle, then use cosine to calculate the angle. If the component represents the adjacent side of the triangle, then use sine to calculate the angle.

$$\boldsymbol{F}_z = 30\,\text{N}\sin 33° = 16.34\,\text{N}$$

$$\boldsymbol{F}_y = 30\,\text{N}\sin 45° = 15\sqrt{2}\,\text{N}$$

$$\boldsymbol{F}_x = 30\,\text{N}\cos 45° = 15\sqrt{2}\,\text{N}$$

2.2. 2D FORCE REPRESENTATION AND VECTOR ADDITION

Problem Author: Tess Wootton-Klebanoff

Type of Problem:
Forces (2D and 3D representation); vector addition and resultant force

What concept is this covering in Statics?
By working through this problem, students will learn vector addition and how it can be applied to real-world applications.

Problem Statement:
You are fighting over the cool new RoboToy XTreme with your sister. While pulling at it in opposite directions, a bird flies down and tries to snatch it away! Your sister, who is not paying any attention, continues to pull on the toy just as before—with a force of 85 N directly away from you (and parallel to the ground). The bird pulls on the toy with a force of 45 N at an angle of 60° above the horizontal, and perpendicular to your sister (see diagram).

With what minimum force must you pull in order to successfully capture the RoboToy XTreme and claim it as your own? (State your answer in vector notation).

[Hint: assume your sister is pulling in the $+x$-direction and the bird is pulling in the direction of the line $z = \sqrt{3}y$.]

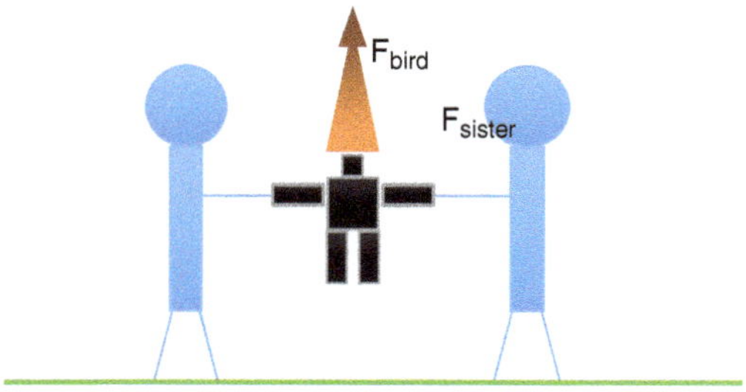

Front view

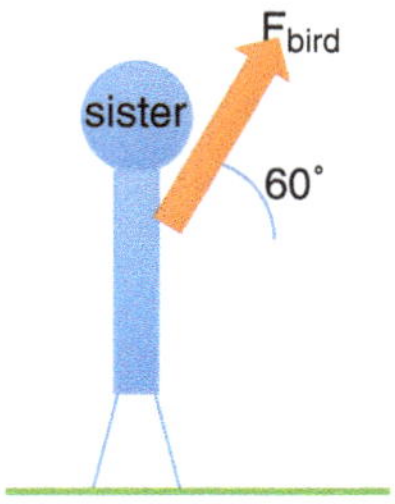

Side view

Goal:
To find the magnitude and direction with which you must pull to be greater than or equal to the combined force of the bird and your sister.

Given:
$\boldsymbol{F}_{\text{sister}} = 85$ N, $\boldsymbol{F}_{\text{bird}} = 45$ N, $\boldsymbol{F}_{\text{sister}}$ acts in the direction of the positive x-axis, $\boldsymbol{F}_{\text{bird}}$ acts in the direction of $z = \sqrt{3}y$.

Assumptions:
See hint.

Equations Needed:
$|\boldsymbol{F}| = \sqrt{\boldsymbol{F}_x^{\,2} + \boldsymbol{F}_y^{\,2} + \boldsymbol{F}_z^{\,2}}$, $\boldsymbol{F}_x = \boldsymbol{F}\cos\theta$, $\boldsymbol{F}_y = \boldsymbol{F}\sin\theta$

Answer:
$\langle -85,\ -22.5,\ -39.87 \rangle$

Solution:
1) Draw x, y, and z axes and put a point (the toy) at the origin.
2) Since the hint tells us the sister is pulling in the direction of the positive x-axis, draw her force as a vector on your axes, starting at the toy. (*Note:* her force can be put anywhere as long as the bird's force is perpendicular to hers in one plane and 60° above hers in another.)

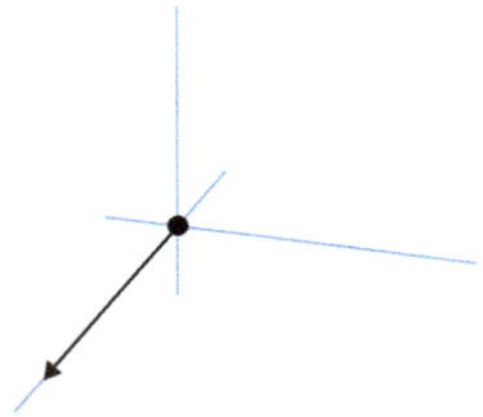

3) Write your sister's force in vector notation:

$$\boldsymbol{F}_{\text{sister}} = \langle 85,\ 0,\ 0 \rangle$$

4) Draw the bird's force on your axes starting at the toy, as shown:

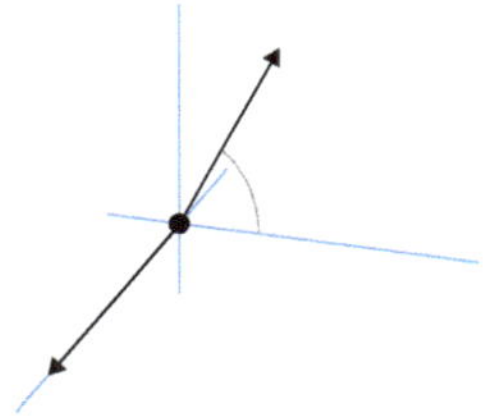

5) Using trigonometry, solve for the components of $\boldsymbol{F}_{\text{bird}}$ and write $\boldsymbol{F}_{\text{bird}}$ in vector notation.
6) Add $\boldsymbol{F}_{\text{sister}} + \boldsymbol{F}_{\text{bird}}$ to solve for the force you must overcome to pull the toy toward yourself:

$$\boldsymbol{F}_{\text{sister}} + \boldsymbol{F}_{\text{bird}} = \langle 85,\ 22.5,\ 39.87 \rangle$$

7) Because you must pull in the opposite direction of this force, change the signs to negative:

$$\boldsymbol{F}_{\text{you}} = \langle -85, -22.5, -39.87 \rangle$$

2.3. MULTIPLE VECTOR, VECTOR ADDITION

Problem Author: Jonathan Shulgach

Type of Problem:
Forces (2D and 3D representation), vector addition

Problem Statement:
In a classroom teamwork activity, a student team must work together to pull a large box across the finish line. The finish line is 800 in. from the origin starting point along the horizontal axis. If each of the four team members pull the box in the directions as shown, how many inches must student a pull the box for the team to cross the finish line?

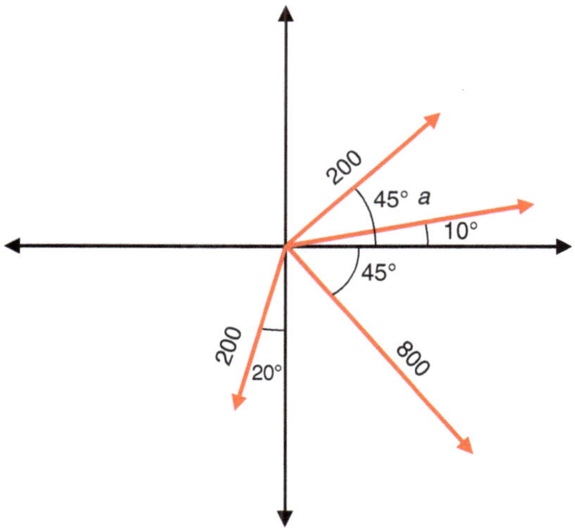

Goal:
Find the distance a needed for the sum of the distances in the x-axis to equal 800 in.

Given:
Diagram

Assumptions:
Assume friction is negligent, forces shown are along the x–y plane.

Equations Needed:

$$\boldsymbol{F}_{\text{Total}} = \sum \boldsymbol{F}_x = \boldsymbol{F}_{x_1,200} - \boldsymbol{F}_{x_2,200} + \boldsymbol{F}_{x_3,800} + \boldsymbol{F}_{x_4,a}$$

Answer:

To find the distance that student a needs to travel in order for the box to travel 800 in., we can set the sum of the forces in the x-direction equal to the total distance traveled.

$$\boldsymbol{F}_{\text{Total}} = \sum \boldsymbol{F}_x = \boldsymbol{F}_{x_1,200} - \boldsymbol{F}_{x_2,200} + \boldsymbol{F}_{x_3,800} + \boldsymbol{F}_{x_4,a}$$

Since each student is pulling at an angle, we must use the x-directional component of each vector. For x_1 the angle is at 45° from the x-axis, so the x-directional trigonometric identity is with the cosine of that angle.

$$\cos 20° = \frac{\boldsymbol{F}_{x_1,200}}{200}$$

Rearranging this to solve for the x-component gives us

$$\boldsymbol{F}_{x_1,200} = 200\cos 45°$$

The same method can be done for $\boldsymbol{F}_{x_4,a}$ and $\boldsymbol{F}_{x_3,800}$

$$\boldsymbol{F}_{x_4,a} = a\cos 20°$$

$$\boldsymbol{F}_{x_3,800} = 800\cos 45°$$

The x-directional trigonometric identity for $\boldsymbol{F}_{x_2,200}$ is the sine of 20°

$$\sin 20° = \frac{\boldsymbol{F}_{x_2,200}}{200}$$

Sum all of the forces in the x-direction

$$800 = 200\cos 45° + a\cos 20° + 800\cos 45° - 200\sin 20°$$

Simplify the equation and solve for a

$$800 = 141.42 + 0.94a + 565.69 - 68.40$$

$$161.29 = 0.94a$$

$$a = 171.59$$

To move the box 800 in. along the x-axis, student a will have to travel 171.59 in.

2.4. 3D VECTOR ADDITION

Problem Author: Yangling Zhou

Type of Problem:

Resultant force and cosine angles

Know how to resolve the forces in Cartesian coordinate system, how to sum forces for resultant components, and how to calculate the cosine angles and magnitude of the resultant.

Problem Statement:

Three forces are acting at point O. ($\boldsymbol{F}_3$ is on x–y plane.)

1) Determine the vector form of the resultant force.
2) Determine the magnitude and cosine angles of the resultant force.

Drawing:

If possible, please take an original photograph or draw as clear as possible the problem. Include any necessary constraints.

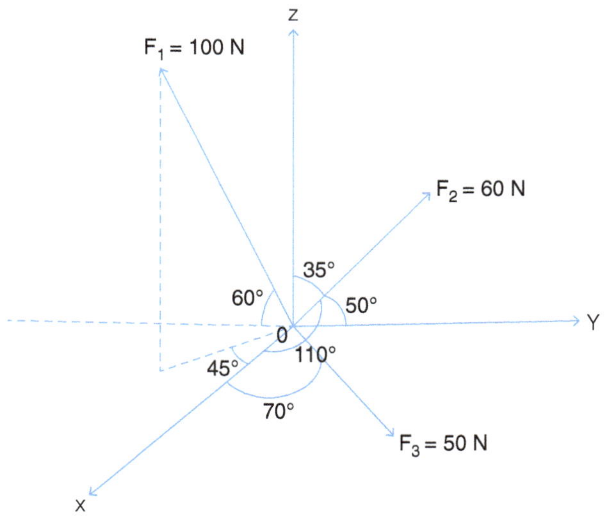

Solution:

$$\vec{F}_R = 31.935\vec{i} + 50.197\vec{j} + 135.752\vec{k}(\text{N})$$

$$|\vec{F}_R| = 148.217\,\text{N}$$

$$\alpha = 77.56°$$

$$\beta = 70.20°$$

$$\gamma = 23.67°$$

Goal:

Find $\vec{F}_R, |\vec{F}_R|, \alpha, \beta, \gamma$

Given:

F_1, F_2, F_3 and angles

Equations Needed:

$$\sum \overline{F_R} = \overline{F_1} + \overline{F_2} + \overline{F_3}$$

$$\cos\alpha = \frac{F_x}{F}, \cos\beta = \frac{F_y}{F}, \cos\beta = \frac{F_z}{F}$$

Fill the blank:

1) Find X, Y, Z components of each force

$F_{1x} =$	$F_{1y} =$	$F_{1z} =$
$F_{2x} =$	$F_{2y} =$	$F_{2z} =$
$F_{3x} =$	$F_{3y} =$	$F_{3z} =$

2) Find X, Y, Z component of the resultant force

$F_{Rx} =$

$F_{Ry} =$

$F_{Ry} =$

$F_R =$

3) Find $|\vec{F}|, \alpha, \beta, \gamma$

Answer:

1) Find X, Y, Z components of each force

$$F_{1x} = 100 \cdot \cos 60° \cdot \sin 45° = 35.355\,\text{N}$$

$$F_{1y} = -100 \cdot \cos 60° \cdot \cos 45° = -35.355\,\text{N}$$

$$F_{1z} = 100 \cdot \sin 60° = 86.603\,\text{N}$$

$$F_{2x} = 60 \cdot \cos 110° = -20.521\,\text{N}$$

$$F_{2y} = 60 \cdot \cos 50° = 38.567\,\text{N}$$

$$F_{2z} = 60 \cdot \cos 35° = 49.149\,\text{N}$$

$$F_{3x} = 50 \cdot \cos 70° = 17.101\,\text{N}$$

$$F_{3y} = 50 \cdot \sin 70° = 46.985\,\text{N}$$

$$F_{3z} = 0\,\text{N}$$

2) Find X, Y, Z component of the resultant force

$$F_{Rx} = F_{1x} + F_{2x} + F_{3x} = 35.355 - 20.521 + 17.101 = 31.935\,\text{N}$$

$$F_{Ry} = F_{1y} + F_{2y} + F_{3y} = -35.355 + 38.567 + 46.985 = 50.197\,\text{N}$$

$$F_{Rz} = 86.603 + 49.149 + 0 = 135.752\,\text{N}$$

$$\vec{F}_R = 31.935\vec{i} + 50.197\vec{j} + 135.752\vec{k}\,(\text{N}) \qquad \text{Ans.}$$

3) Find $|\vec{F}_R|, \alpha, \beta, \gamma$

$$|\vec{F}_R| = \sqrt{31.935^2 + 50.197^2 + 135.752^2} = 148.217\,\text{N} \qquad \text{Ans.}$$

$$\cos\alpha = \frac{F_{Rx}}{F_R} = 0.2154 \Rightarrow \alpha = 77.56° \qquad \text{Ans.}$$

$$\cos\beta = \frac{F_{Ry}}{F_R} = 0.3387 \Rightarrow \beta = 70.20° \qquad \text{Ans.}$$

$$\cos\gamma = \frac{F_{Rz}}{F_R} = 0.9159 \Rightarrow \gamma = 23.67° \qquad \text{Ans.}$$

2.5. UNIT VECTORS ON A 3D OBJECTED

Problem Author: Sarah Hemler

Type of Problem:

Through this problem, students should be able to solve for 3D unit vector components using geometry.

Problem Statement:

The bungee of a folding luggage cart is connected to two locations on the cart. Find the 3D unit vectors $\widehat{\boldsymbol{u}_{OA}}$, $\widehat{\boldsymbol{u}_{OB}}$. If the tension is a uniform force of 40 lb. along the bungee, solve for the force vectors, $\overrightarrow{\boldsymbol{F}_{OA}}$ and $\overrightarrow{\boldsymbol{F}_{OB}}$.

Drawing:

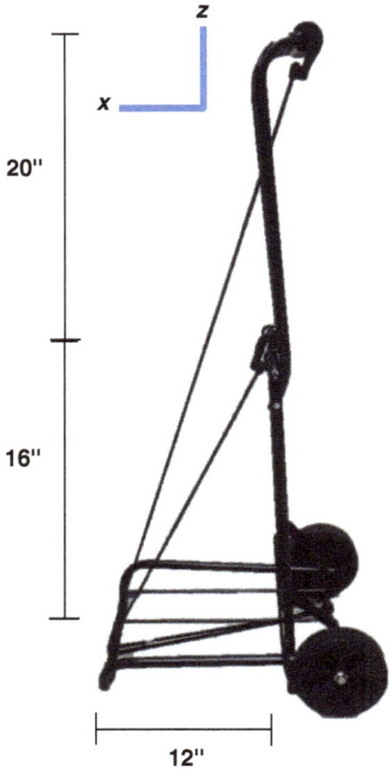

Courtesy of Sarah Hemler

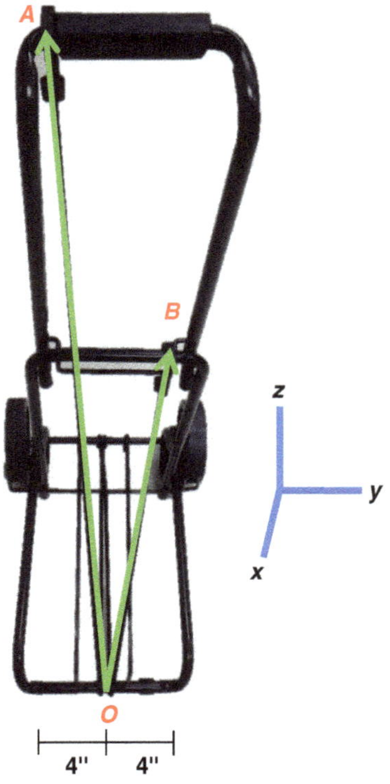

Solution:

$\widehat{\boldsymbol{u}_{OA}} = -0.314, -0.105, 0.943$ $\widehat{\boldsymbol{u}_{OB}} = -0.588, 0.196, 0.784$

$\overrightarrow{\boldsymbol{F}_{OA}} = -12.36, -4.20, 37.73\,\text{lb.}$ $\overrightarrow{\boldsymbol{F}_{OB}} = -23.52, 7.84, 31.36\,\text{lb.}$

Goal:

Find the 3D unit vector components for $\overrightarrow{\boldsymbol{OA}}$ and $\overrightarrow{\boldsymbol{OB}}$:

Given:

Geometry

Assumptions:

Bungee is static. Coordinate system is established.

Equations Needed:

$$\boldsymbol{r} = \boldsymbol{x}, \boldsymbol{y}, \boldsymbol{z} \qquad |\vec{r}| = \sqrt{r_x{}^2 + r_y{}^2 + r_z{}^2} \qquad \hat{\boldsymbol{u}} = \frac{\vec{\boldsymbol{r}}}{|\vec{r}|} \qquad \vec{\boldsymbol{F}} = \boldsymbol{F} \times \hat{\boldsymbol{u}}$$

Answer:

Locate points A and B relative to O. Find the position vectors for $\overrightarrow{\boldsymbol{OA}}$ *and* $\overrightarrow{\boldsymbol{OB}}$:

$$\boldsymbol{O} = (0,0,0), \qquad \boldsymbol{A} = (-12,-4,\ 36), \qquad \boldsymbol{B} = (-12,\ 4,\ 16)$$

$$\overrightarrow{r_{OA}} = -12,-4,\ 36 \text{ in.}$$

$$\overrightarrow{r_{OB}} = -12,\ 4,\ 16 \text{ in.}$$

Find the magnitude of the two vectors:

$$\left|\overrightarrow{r_{OA}}\right| = \sqrt{(-12)^2 + (-4)^2 + (36)^2} = \sqrt{1456} = 4\sqrt{91} = 38.16 \text{ in.}$$

$$\left|\overrightarrow{r_{OB}}\right| = \sqrt{(-12)^2 + (4)^2 + (16)^2} = \sqrt{416} = 4\sqrt{26} = 20.40 \text{ in.}$$

Solve for the unit vectors:

$$\widehat{u_{OA}} = \frac{\overrightarrow{r_{OA}}}{\left|\overrightarrow{r_{OA}}\right|} = \frac{-12,-4,\ 36 \text{ in.}}{4\sqrt{91} \text{ in.}} = -\frac{3}{\sqrt{91}}, \frac{-1}{\sqrt{91}}, \frac{12}{\sqrt{91}} = -0.314, -0.105, 0.943$$

$$\widehat{u_{OB}} = \frac{\overrightarrow{r_{OB}}}{\left|\overrightarrow{r_{OB}}\right|} = \frac{-12,\ 4,\ 16 \text{ in.}}{4\sqrt{26} \text{ in.}} = -\frac{3}{\sqrt{26}}, \frac{1}{\sqrt{26}}, \frac{4}{\sqrt{26}} = -0.588, 0.196, 0.784$$

Solve for the force vectors:

$$\overrightarrow{F_{OA}} = \boldsymbol{F} \times \widehat{u_{OA}} = 40 \text{ lb.} \times -0.314, -0.105,\ 0.943 = -12.36, -4.20, 37.73 \text{ lb.}$$

$$\overrightarrow{F_{OB}} = \boldsymbol{F} \times \widehat{u_{OB}} = 40 \text{ lb.} \times -0.588,\ 0.196,\ 0.784 = -23.52, 7.84, 31.36 \text{ lb.}$$

2.6. FINDING AN ANGLE BETWEEN TWO VECTORS

Problem Author: Jordyn Schroeder

Type of Problem:
Dot Product
Students should be able to find the angle between two vectors, on a 2D axis.

Problem Statement:
A side view of a lawn chair is shown in the following. Find the angle of incline of the chair, θ.

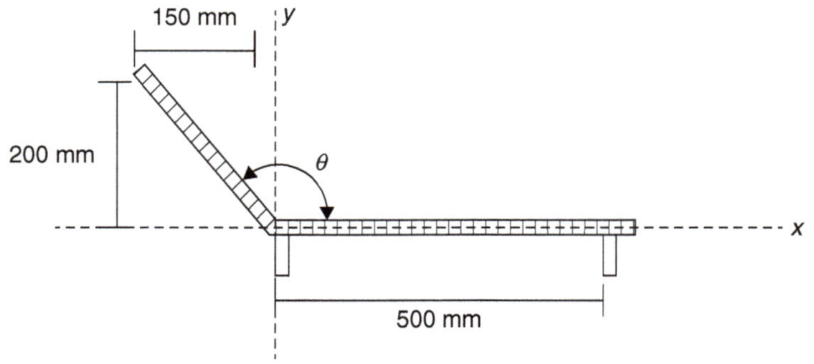

Goal:
Find the angle between two vectors.

Given:
X and Y components of vectors 1 and 2 <–150, 200> mm and <500, 0> mm, respectively.

Assumptions:
N/A

Equations Needed:

$$A \cdot B = AB\cos(\theta) = AxBx + AyBy + AzBz$$

$$|A| = \sqrt{(Ax^2 + Ay^2 + Az^2)}$$

Answer:

$$|A| = \sqrt{(Ax^2 + Ay^2 + Az^2)} = \sqrt{(-150\,\text{mm})^2 + 200\,\text{mm}^2} = 250\,\text{mm}$$

Using $A \cdot B = AB\cos(\theta) = AxBx + AyBy + AzBz \rightarrow 250 \times 500\cos(\theta) = -150 \times 500 + 200 \times 0 \rightarrow$ $\theta = 126.9°$.

CHAPTER 3

Moments (2D and 3D Representation)

Sometimes it just takes a couple to make a moment

A. MOMENT DEFINITION: MORE THAN JUST AN INSTANT IN TIME

Moments, also known as torque, are a type of "force" that causes an object to want to bend, twist, or rotate.

Moments have units of

- Newton meters (N·m) in SI units
- Pound feet (lb. × ft.) in US Customary units

What Does a Moment Feel Like?

Try holding your pencil or a ruler from an end and then hold it in the center as shown. What do you notice?

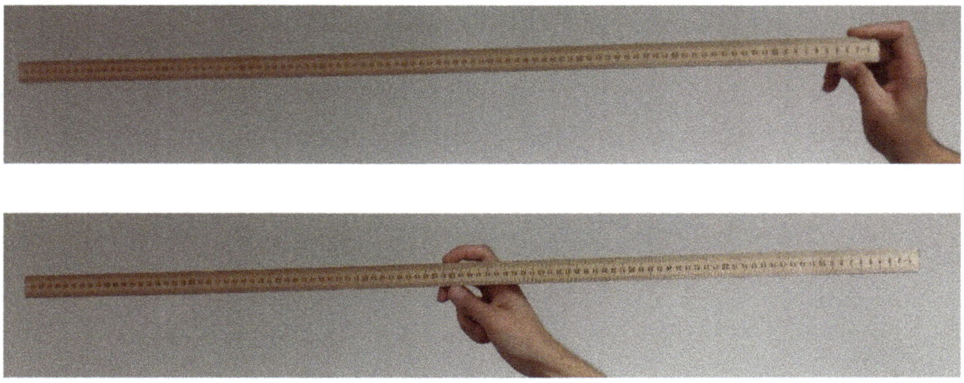

Courtesy of Jamie Gurganus

Do you find it is more difficult to hold it from the end? This happens because you must create a moment to hold up the weight. Moments help to keep things in equilibrium in statics or provide rotation to drive shafts, for example, in dynamics.

Some common examples of the application of a moment include closing a door, tightening a nut, and so on.

HOW ARE MOMENTS CREATED (POSITION VECTOR *r*)?

Moments are created by a force (vector) that is offset from a point located a distance away from the force.

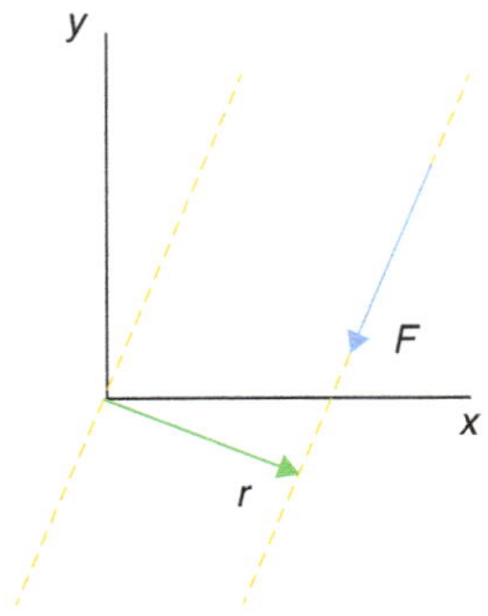

The offset of a force is defined as the position vector, r. A position vector is a vector of magnitude and direction with units of length. Given 2 points along a line A and B, the position vector can be calculated by simply subtracting the coordinates of B from the coordinates of A.

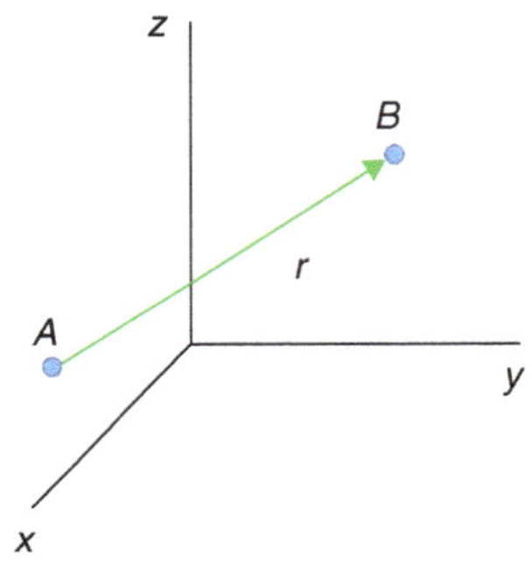

$$r = (B_x - A_x)i + (B_y - A_y)j + (B_z - A_z)k$$

MOMENTS IN 2D (SCALAR) AND 3D (VECTOR)

The moment center is the point (2D) or axis (3D) about which the moment is acting. In 2D, the moment can act either negative direction (rotating clockwise) or positive direction (rotating counterclockwise).

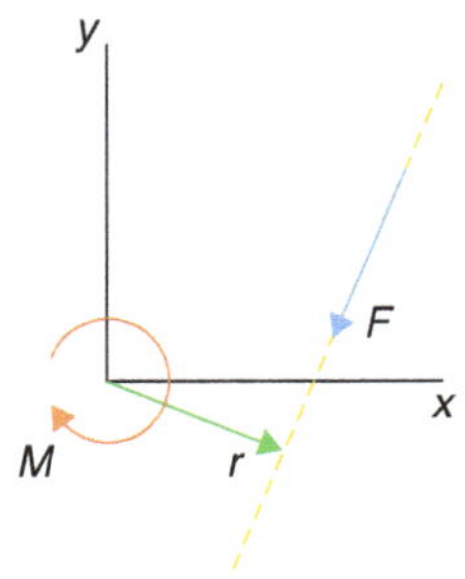

This moment is acting in negative Z axis (into the page).

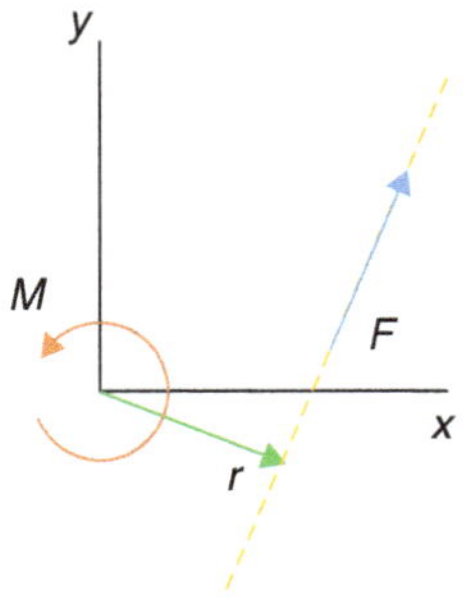

This moment is acting in positive Z axis (toward you).

In 3D, it gets a little more complicated, but don't fret! There is a super cool method for determining if a moment is positive or negative, and it is called the right-hand rule (RHR).

Right-Hand Rule

The RHR states that if you curl your fingers in the direction that the moment is acting, your thumb will point outward indicating the direction of the moment (in this case, out of the page). In this case, the moment is positive with the thumb pointed in the positive direction as shown below.

Courtesy of Jamie Gurganus

The following figures show when the moment created is negative (going into the page):

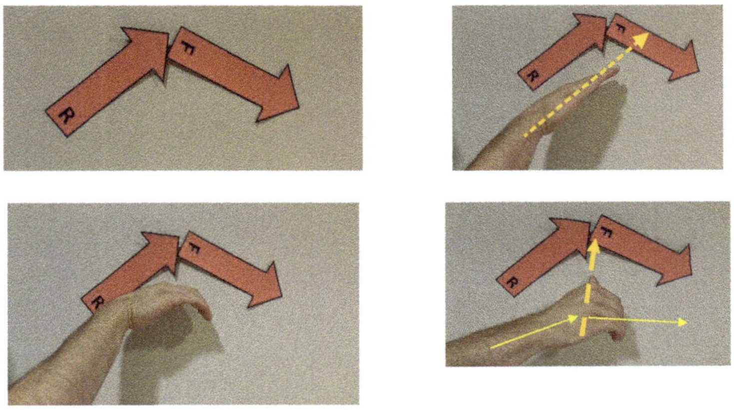

Courtesy of Jamie Gurganus

B. MATHEMATICAL REPRESENTATION OF MOMENTS

Just like forces, moments have both magnitude and direction, which means they are yet another type of vector.

Each component of the moment represents *how much* the object wants to bend, twist, or rotate along the respective axis. As stated earlier, moments are created by a force and some offset.

Scalar Calculation (For Two Dimensions)

As a scalar, moments are represented by the force magnitude multiplied by the perpendicular distance of the force to the moment center.

$$||\boldsymbol{M}|| = ||\boldsymbol{F}||\,||\boldsymbol{r}||$$

$$||\boldsymbol{M}|| = ||\boldsymbol{r}||(||\boldsymbol{F}||\sin\theta)$$

θ is the angle between these two vectors when they are placed tail to tail such that $0° < \theta < 180°$

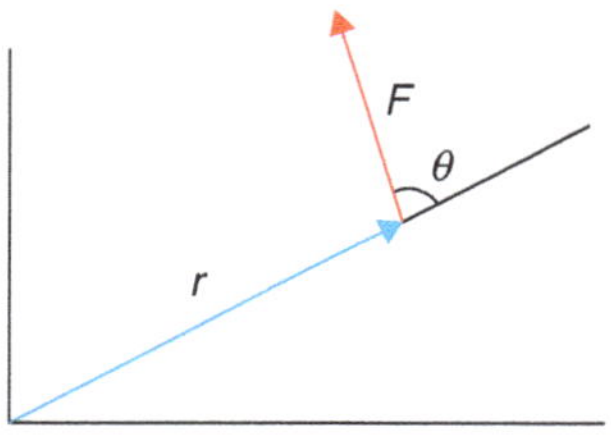

Vector Calculation (for Two or Three Dimensions)

To analyze a moment as a vector we can do one of two methods, scalar or vector.

For simple cases, especially 2D, we can perform the *scalar analysis* described previously for each component.

For more complex (3D especially) cases, it's easier to use the cross product, which is how two vectors are multiplied together.

The cross product is defined as the position vector r "crossed" with the force vector F.

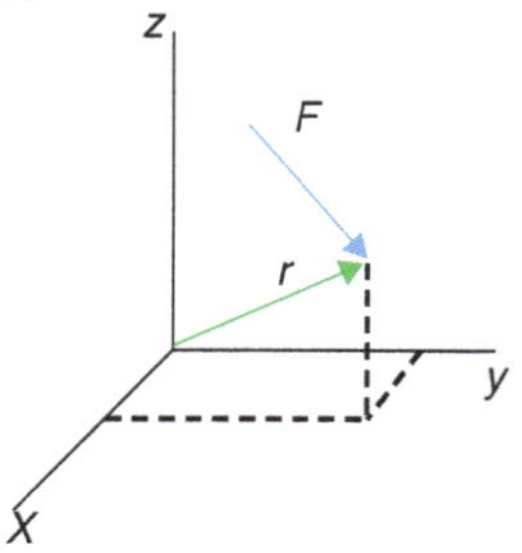

$$M = r \times F$$

Cross products are evaluated using the determinant

$$M = \begin{vmatrix} i & j & k \\ r_x & r_y & r_z \\ F_x & F_y & F_z \end{vmatrix} = Mx + My + Mz$$

$$M = (r_y\ F_Z - r_Z\ F_y)i - (r_x F_z - r_z F_x)j + (r_x F_y - r_y F_x)k$$

Note: Helpful hints for solving determinants can be found in the Appendix A.

When solving the determinant, unlike with dot products, the result of the cross product is a *vector*. So once you solve the determinant, you already have the components (i, j, k). As with any vector, its magnitude is found by simply using the Pythagorean theorem.

DIRECTION COSINES AND THE UNIT VECTOR ... FOR MOMENTS TOO!

The direction of the moment can be defined in terms of the directional cosines:

$$\cos\theta_x = \frac{M_x}{||M||}$$

$$\cos\theta_y = \frac{M_y}{||M||}$$

$$\cos\theta_z = \frac{M_z}{||M||}$$

The moment's direction can also be described by a unit vector u:

$$u = \cos\theta_x i + \cos\theta_y j + \cos\theta_z k$$

FINDING MOMENTS IN A PARTICULAR DIRECTION

To find a moment acting in a particular direction, such as along a particular axis or object, we must utilize unit vectors.

Recall finding force projections from Chapter 2. To find the projections, we use the force vector and "dot" it with the unit vector in a particular direction. This method is additionally used for moments in a particular direction.

In the following equation, we are solving for the projection (the shadow) of the moment M in the direction of the unit vector u_a (see in the figure to the right).

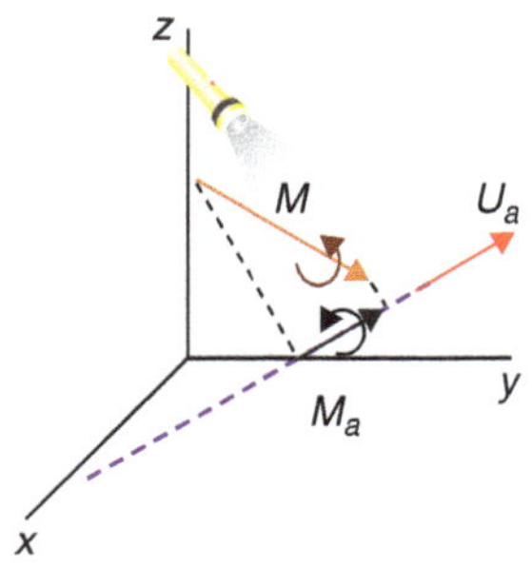

Vector: Multiply the *magnitude* with the unit vector of the line (u_a)

$$M_a = M \cdot u_a$$

Angle: To find the angle between u_a and M you can use the dot product rule

$$\cos\theta = \frac{M_x u_{ax} + M_y u_{ay} + M_z u_{az}}{||M||\ ||u_a||}$$

This is useful as some materials fail in certain directions caused by force moments acting in another direction.

COUPLES

If you have ever turned a wheel with your hands, such as a steering wheel in a car, then you have probably created a force couple … that is, unless you are the kind of person who drives with one hand at the top of the steering wheel.

Do It Yourself With Couple Moments

Hold your pencil at each end and, using your fingers, apply forces perpendicular to the length of the pencil. If the forces you apply are opposite one another, then the pencil will not move side to side (or up and down), instead it will rotate about its center (the moment center). When two forces simplify into a single moment, they are called a couple.

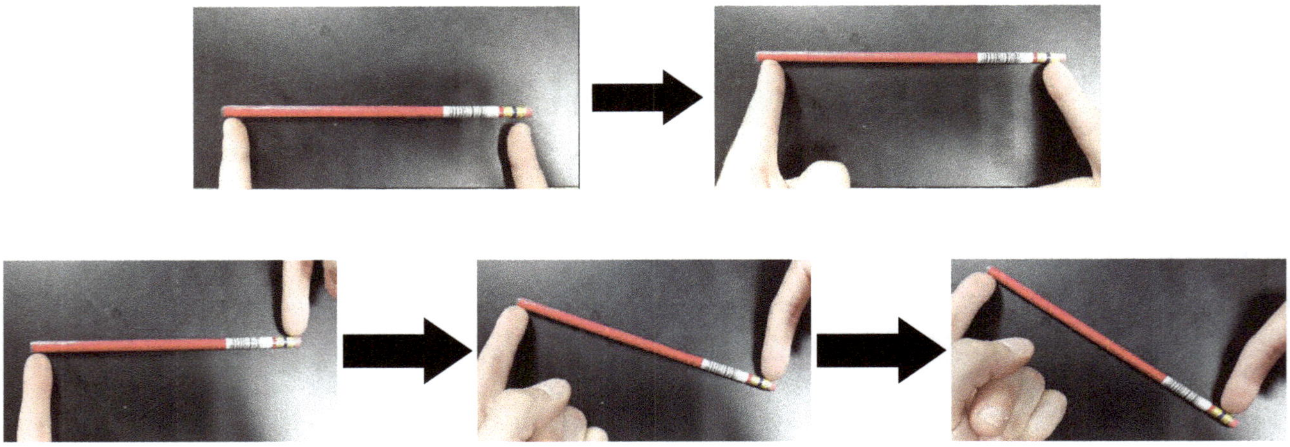
Courtesy of Jamie Gurganus

STEPS TO SOLVE FOR A MOMENT

Scalar Moments (2D)

For scalar moments, there are three methods which you can solve a problem.

$$||\boldsymbol{M}||=||\boldsymbol{F}||\ ||\boldsymbol{r}||$$

Method One:

Find the perpendicular distance, $r_{\perp}$, between the force and the moment center. Then, given the magnitude of the force, use the preceding equation(s).

Method Two:

Solve for the x and y components (as shown in the following) of the force and the position vector, then multiply by the y and x distances, respectively.

$$M = (r_x F_y + r_y F_x)$$

Method Three:

Find the position vector ***r***, then use the scalar form of the cross product equation.

$$||\boldsymbol{M}|| = ||\boldsymbol{r}|| \, (||\boldsymbol{F}|| \sin\theta)$$

Vector Moments (3D)

For vector moments, there are two methods to solve for moments.

Method One:

Find the components of the force (x, y, z) and multiply them with the perpendicular distances from the moment center in each direction.

$$M = (r_y F_z + r_z F_y)i + (r_x F_z + r_z F_x)j + (r_x F_y + r_y F_x)k$$

Note: Notice, in the above equation, the signs are all positive. You will need to use the RHR to determine the direction of the moment (whether it's positive or negative).

Method Two:

Find the position vector ***r***, then use the vector form of the cross product equation.

$$M = r \times F$$

$$M = \begin{vmatrix} i & j & k \\ r_x & r_y & r_z \\ F_x & F_y & F_z \end{vmatrix} = Mx + My + Mz$$

SECTION 3

Problems

3.1. EQUILIBRIUM OF A SEESAW

Type of Problem:

This problem covers the concept of moments. The student should be able to use across product to find the moment about a point.

Problem Statement:

Two children sit on a seesaw at an angle of 20°. The first child, having a mass of 45 kg, sits 1.2 m from center O while the second child sits 1 m away from center O. Determine the mass of the second child.

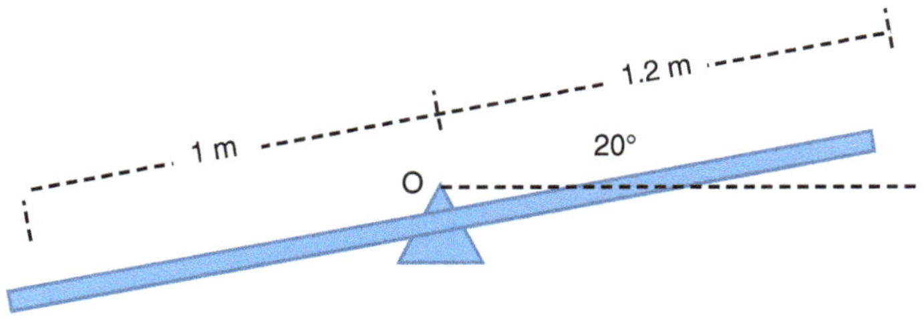

Goal:

Find the mass of the second child, m_2

Given:

The mass of the first child, m_1
The angle of the seesaw, 20°
The length of the seesaw

Equations Needed:

$$M = r \times F$$

Answer:

$$\overrightarrow{OA} = 1.2\cos 20,\ 1.2\sin 20$$

$$\overrightarrow{OB} = -1\cos 20,\ -1\sin 20$$

$$M_1 = \overrightarrow{OA} \times m_1 g = \begin{vmatrix} i & j & k \\ 1.128 & 0.410 & 0 \\ 0 & -441.45 & 0 \end{vmatrix}$$

$$M_1 = (-497.95\,\text{N}\cdot\text{m})k$$

$$\sum M = 0 = M_1 + M_2$$

$$M_1 = -M_2$$

$$M_2 = \overrightarrow{OB} \times m_2 g = \begin{vmatrix} i & j & k \\ -0.94 & -0.342 & 0 \\ 0 & -m_2 g & 0 \end{vmatrix}$$

$$-497.95k = [(-0.94)(-9.81)m_2]k$$

$$M_2 = 54\,\text{kg}$$

3.2. APPLIED MOMENT (TORQUE) FROM A WRENCH

Problem Author: Jordyn Schroeder

Type of Problem:

2D moments

Students should be able to understand moment (torque) on a 2D axis.

Problem Statement:

Sarah has a flat tire and needs to change it. The bolts on the tire require a torque of 30 N·m to loosen. What is the minimum force, F, that Sarah must apply to the end of the 50 cm wrench at an angle, θ, of 65°.

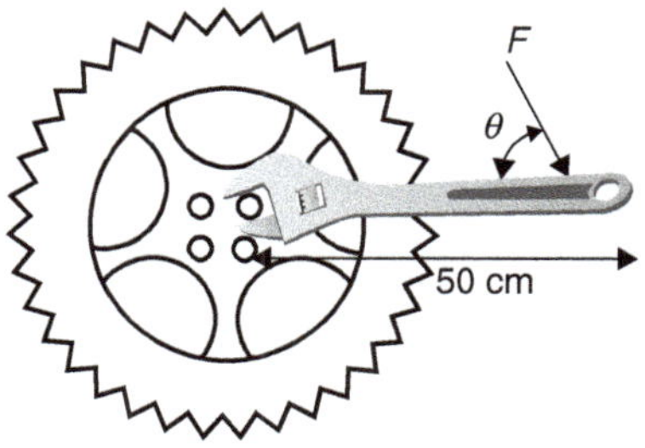

Goal:

Find force, F, that creates a moment on the bolt of 30 N·m.

Given:

Torque t = 30 N·m, θ = 65°, distance d = 50 cm

Assumptions:

N/A

Equations Needed:

$$M = t = F \times d$$

Answer:

Using $M = t = F \times d \rightarrow 30 = F\sin(65) \times 50 \text{ m} \rightarrow F = 66.2 \text{ N}$

3.3. SEESAW BALANCE #2

Question 1:

This question will help students understand the concept of moments and will allow them to see that even a small force at a large distance will create a large moment.

"Equivalent Moments and force"

Problem Statement:

Zach is a typical 180 lb. mechanical engineering student. If Zach is sitting on a seesaw, at distance 4 ft. from its fulcrum, how far away would his 1-lb. statics textbook have to be in order to balance out his weight?

Drawing:

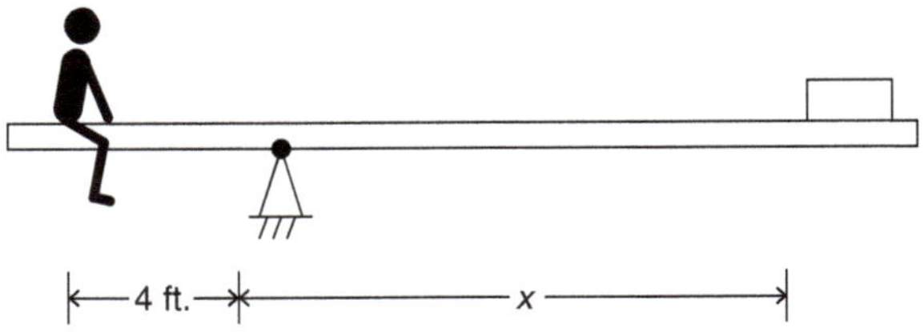

Find:

The length of the seesaw in order to reach equilibrium.

Given:

Length of one side, mass of Zach, and mass of textbook.

Assumptions:

The seesaw will be in equilibrium; the mass of the book and the mass of Zach are uniform; gravitational acceleration is 32.2 ft./s^2.

Equations Needed:

Sum of the moments about O.

Solution:

$$F\text{Zach} = 180\ \text{lbm} \times 32.2\ \frac{\text{ft.}}{\text{s}^2} = 5{,}796\ \text{N}$$

$$F\text{book} = 1 \text{ lbm} \times 32.2 \frac{\text{ft.}}{\text{s}^2} = 32.2 \text{ N}$$

$$\sum M_O = 0 = 5{,}796 \text{ N} \times (4 \text{ ft.}) - 32.2 \text{ N} \times (x \text{ ft.})$$

$$x = 720 \text{ ft.}$$

FBD:

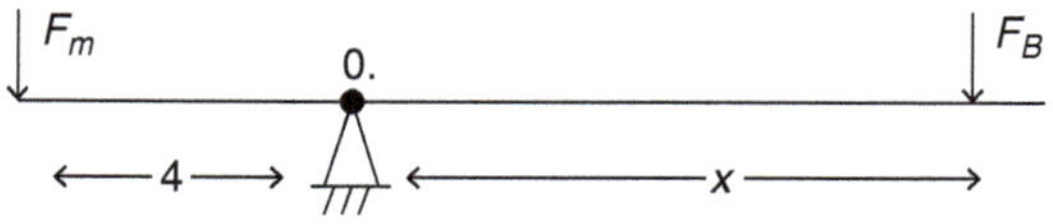

3.4. 3D MOMENT BALANCE ON A TIRE IRON

Problem Author: Neil S. Rothman

Type of Problem:

This problem covers the concept of resolving a force into an equivalent force and moment, specifically using a real-life example of an "L"-shaped tire iron and the loads it applies to the lug nut.

The student should be able to translate the applied force at the end of the wrench into a force and moment (torque) at the lug nut (point O).

Problem Statement:

The following is an "L"-shaped tire iron that is used to tighten/loosen the lug nuts that hold a tire onto a car. Assuming the user applies a force, P, as shown at point B, determine the equivalent force and/or moment applied to the lug nut at O (i.e., the reactions) in terms of the applied load P. You may assume that the wrench is fixed at point O.

Drawing:

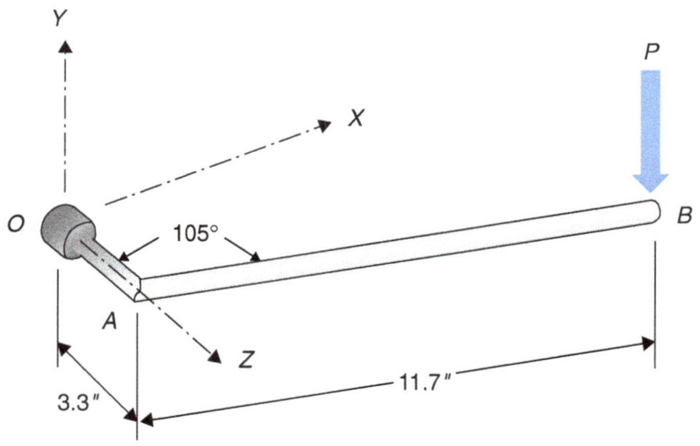

Goal:

Find the reaction forces and moments at point O where the wrench meets the lug nut.

Given:

Wrench dimensions, applied load P.

Assumptions:

The wrench is fixed at O (i.e., there can be both reaction forces and moments at O); since the only applied load is in the Y-direction, there will not be any reaction forces in the X- or Z-directions; we expect moments about both the X and Z axes.

Equations Needed:

$$\sum F_Y = 0$$

$$\sum M_X = 0$$

$$\sum M_Z = 0$$

Answer:

The first step in the solution is to convert the force at B into a force applied on the rotational centerline of the wrench (the Z-axis) and a moment. This is done by placing two equal and opposite forces (for convenience we will call them $P1$ and $P2$, but $P1 = P2 = P$) on the Z-axis an equal distance in the Z-direction as the applied force P from point O, at location A'. The distances OA' and $A'B$ can be calculated using simple trigonometry with the angle given.

Distance $A'B = 11.7 \cos 15 = 11.3''$

Distance $OA' = 3.3 + 11.7 \sin 15 = 6.33''$

See the following figure for this construction.

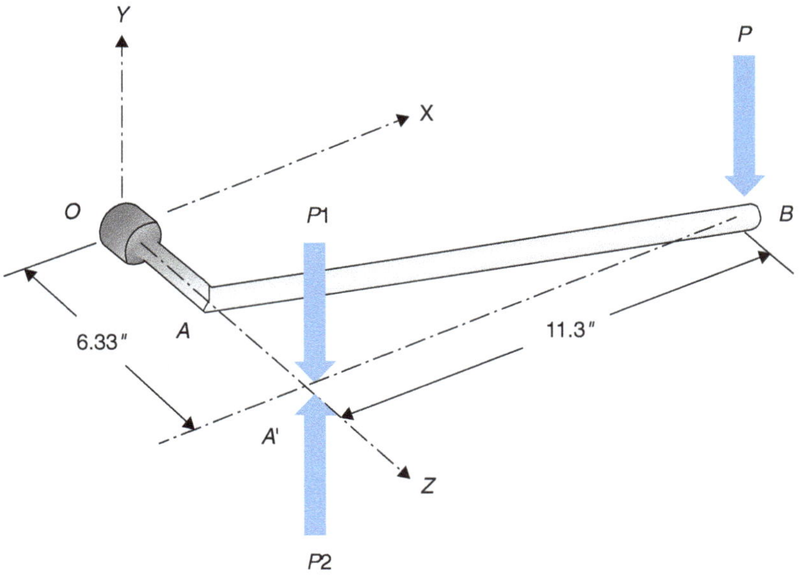

There is now a couple (moment) about the Z-axis created by P and $P2$ which is equal to the force times the separation distance: $M_Z = 11.3\,P$ (remember that $P2 = P$). There is also a force acting in the negative Y-direction ($P1$) which is equal to P. This completes the conversion of the force P at B into a force at A' and a couple about the Z-axis.

We can now draw the free-body diagram (FBD) for the wrench with reaction forces/moments at O.

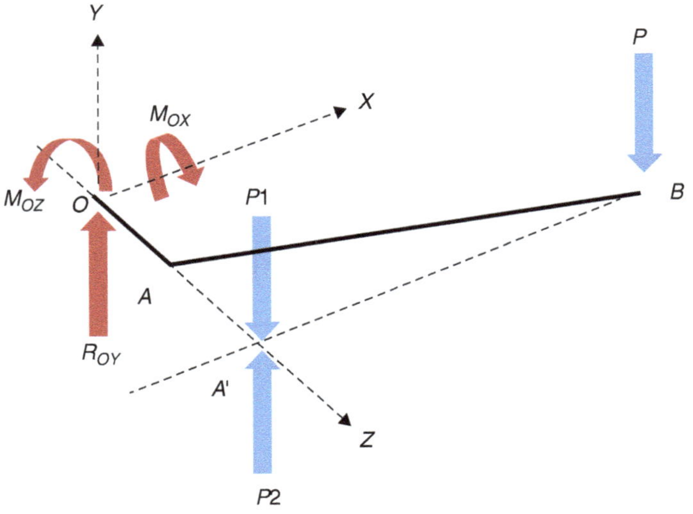

Using summation of forces and moments at point O, we find the following:

$$\sum F_Y = 0 = R_{OY} - P1 \rightarrow R_{OY} = P1 = P$$ Ans

$$\sum M_X = 0 = M_{OX} + 6.33\,P1 \rightarrow M_{OX} = -6.33\,P1 = -6.33\,P$$ Ans

$$\sum M_Z = 0 = M_{OZ} - 11.3\,P \rightarrow M_{OZ} = 11.3\,P$$ Ans

Alternate Solution:

The reaction force ROY is found as above using summation of forces.

The moment at O is found using the vector representation of the applied force and position vector R:

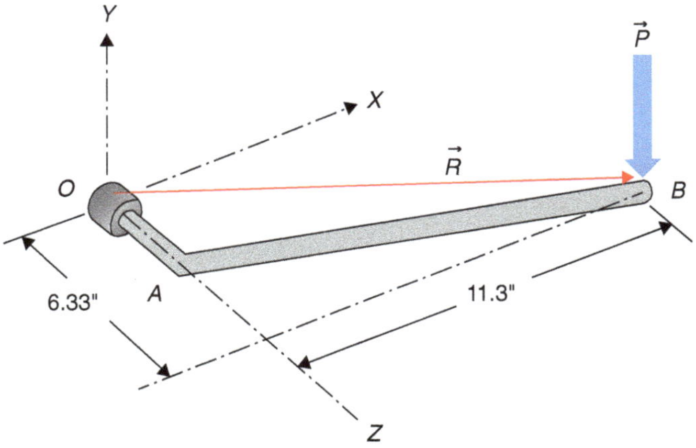

$$\vec{R} = 11.3\vec{i} + 6.33\vec{k}$$

$$\vec{P} = -P\vec{j}$$

$$\vec{M} = \vec{R} \times \vec{P}$$

Writing the cross-product out:

$$\vec{R} \times \vec{P} = \begin{vmatrix} \vec{i} & \vec{j} & \vec{k} \\ 11.3 & 0 & 6.33 \\ 0 & -P & 0 \end{vmatrix}\begin{vmatrix} \vec{i} & \vec{j} \\ 11.3 & 0 \\ 0 & -P \end{vmatrix} = 6.33P\vec{i} - 11.3P\vec{k}$$

which is the same result as the previous method.

3.5. SEESAW BALANCE #3

Problem Author: Sarah Hemler

Type of Problem:

In this problem, students should be able to determine the sum of the moments about a point. This is also a problem in which students must analyze a real-world problem and determine the optimized solution.

Problem Statement:

A father is playing with his 4-year-old son at the park. The father sets his son who weighs 36 lb. on the seesaw 3 ft. from the fulcrum. The father, who weighs 216 lb., then sits 4 ft. from the fulcrum and places his feet on the ground to balance the seesaw. The total length of the symmetrical seesaw is 12 ft. and there are three seats behind the son that are in 1 ft. increments. In a world, where all 4-year olds weigh 36 lb., is it possible for the seesaw to be balanced once the father lifts his feet off the ground? If so, how many 4-year olds are necessary to balance. If not, what would four, 4-year olds need to weigh to balance the seesaw?

Drawing:

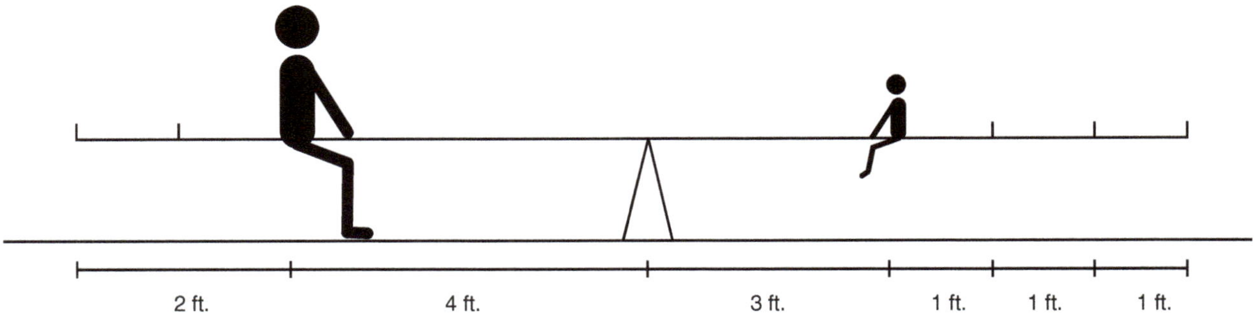

Assumptions:

Fulcrum acts as a pin; weight of individuals acts in the negative vertical direction.

Equations Needed:

$$\sum M_z = 0$$

Answer:

Free-Body Diagram 1:

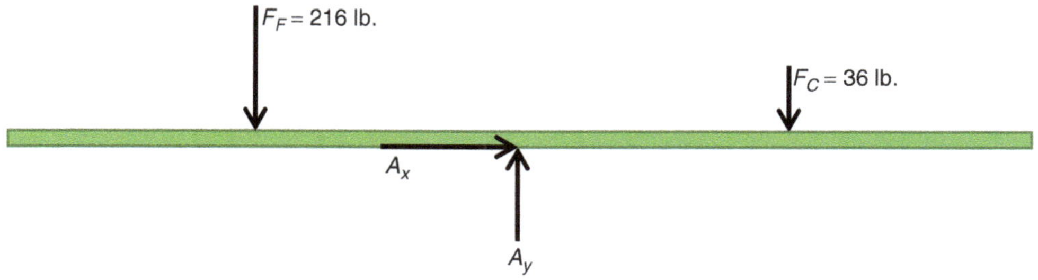

1) Take the moment about the fulcrum (point A) accounting for the father and the child's weights from FBD 1 above

$$\sum M_z = F_F \times 4 \text{ ft.} - F_C \times 3 \text{ ft.} = 216 \text{ lb.} \times 4 \text{ ft.} - 36 \times 3 \text{ ft.} = 756 \text{ lb.} \times \text{ft.} \neq 0$$

Since the moment about the fulcrum is not equal to zero, there will be movement of the seesaw.

2) Therefore, we must calculate if 2, 3, or 4 children weighing 36 lb. each can balance the seesaw with the father. Let's start with the maximum number of children to see if it is even possible to balance.

Free-Body Diagram 2:

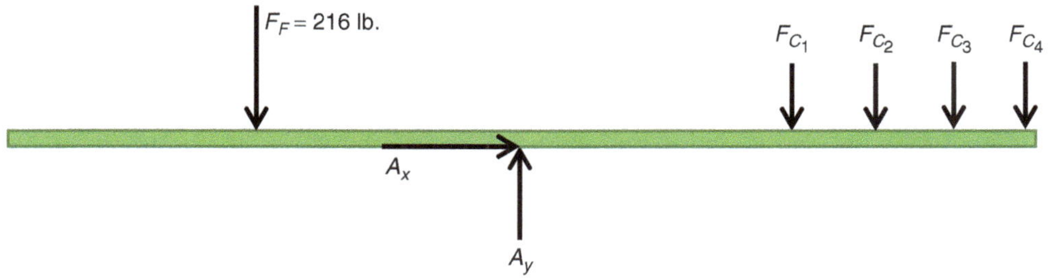

Take the moment about the fulcrum (point A) accounting for the father and the four children's weights from FBD 2 above

$$\sum M_z = F_F \times 4 \text{ ft.} - F_{C_1} \times 3 \text{ ft.} - F_{C_2} \times 4 \text{ ft.} - F_{C_3} \times 5 \text{ ft.} - F_{C_4} \times 6 \text{ ft.} = 291 \text{ lb.} \times \text{ft.}$$

Since the moment about the fulcrum is still not equal to zero, there will be movement of the seesaw with four children weighing 36 lb.

3) Thus, we must calculate the standard weight of four children which will balance the seesaw at the four positions shown in FBD 2. We can use the sum of the moment equation from previous to calculate the weight:

$$F_{C_1} = F_{C_2} = F_{C_3} = F_{C_4} = F_C$$

$$\sum M_Z = F_F \times 4 \text{ ft.} - F_C(3 \text{ ft.} + 4 \text{ ft.} + 5 \text{ ft.} + 6 \text{ ft.}) = 0 \rightarrow F_C = 48 \text{ lb.}$$

Four children must each weight 48 lb. and be positioned as seen in FBD 2 to balance the father on the other side of the seesaw.

CHAPTER 4

Free-Body Diagrams and Analysis

Let's keep it in equilibrium

So, let's get into the nitty gritty of statics. Forces. There are helpful forces and unhelpful forces engineers deal within design applications. The biggest challenge, especially in statics, is to keep our structures in equilibrium.

TYPES OF LOADS ACTING A SYSTEM

For now, let's focus just on loads acting externally on a body. Later, internal forces will be important with frames, machines, trusses, and beams.

Two types of external loads that act on a system include:

A. Across the System: such forces include gravity.
B. Directly on the System: This could include, depending on your boundary, supports such as a bolt, rope, or cable.

FREE BODY DIAGRAM

One of the most fundamental pieces in solving a rigid-body equilibrium problems is knowing how to draw a free-body diagram (FBD). If you don't do this correctly, the rest of your analysis will have problems and your answer will not make sense.

Steps to Drawing an FBD

Note: There are many variations presented on how to draw an FBD. Choose what you are the most comfortable with and DO IT!

1. Identify the object or point (system) that you will be drawing.
2. Identify all the forces acting *directly and across* on the object and the object exerting them. (If the system has weight, make sure to include that too.)

   Note: Do not include forces by an object acting through another object—in other words, only include forces acting directly on the system.

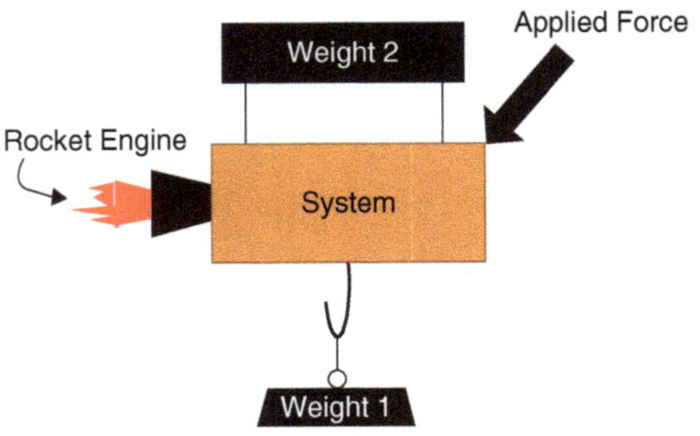

3. Draw a vector to represent each force.
   Draw the vector in the direction the force is being exerted, and label it by (a) the type of force, (b) the object exerting the force, and (c) the object receiving the force (which will be your object of interest).
   Some examples of force labels are:

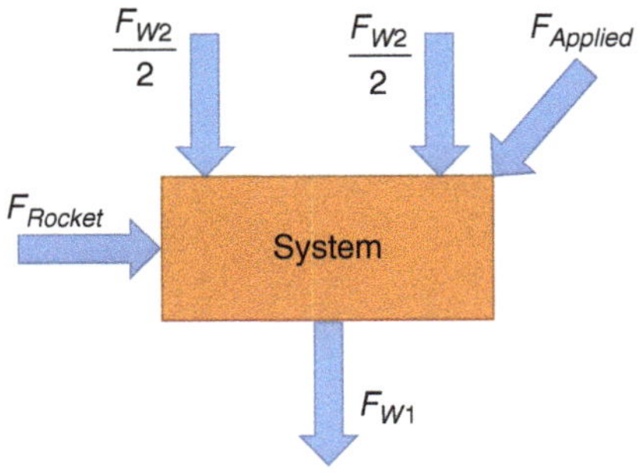

$$F_{\text{exerting object}}$$

$$F_{\text{exerting object} \to \text{system}}$$

   Note: This is often used when you have multiple FBDs so that you can distinguish between them especially when including internal forces, like in frames (next chapter).

$$W_{\text{object}}$$

   Note: W is only used to signify weights.

   Sometimes, P is used to represent a force, but don't be confuse, it still represents a point (P) load.

4. If the object is stationary or is moving at a constant velocity, the vectors (forces) should graphically add up to zero. If the object is accelerating, the sum of the vectors should produce a vector in the same direction as the acceleration.

   Note: If the object is accelerating, we are no longer in static equilibrium.

PLANAR (2D) AND NONPLANAR (3D) FORCES

Recall

Planar: If all the forces can be shown or represented on a single plane and all the moments are acting perpendicular to the plane, than it is planar (2D). If this condition is not satisfied than the system is said to be nonplanar.

Support Reactions: The purpose of the support reactions is to prevent either translation or rotation by imposing an opposite force on the system.

There are many types of support reactions that are used in design. However, the most common discussed in analysis include a pin, roller, and fixed in a system. Some of these, planar, are shown in the following with the reaction forces.

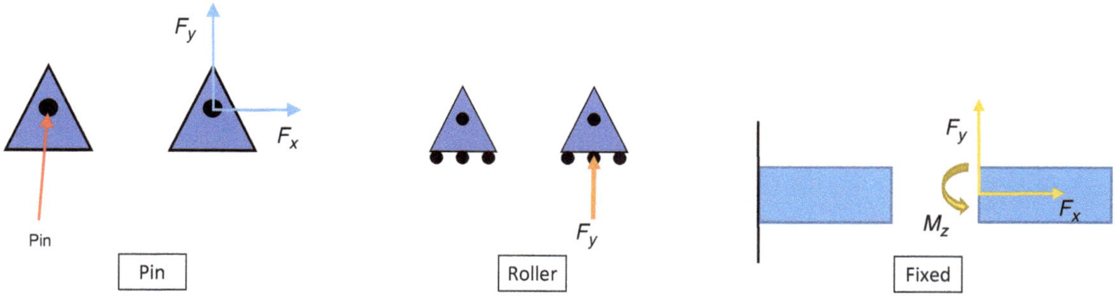

Note: It's advisable for the student to refer to their text books for more supports and their force reactions.

CONDITIONS FOR EQUILIBRIUM

Force Equilibrium Equations (for Force Balance)

Remember to include all the forces acting on the system. It is easier to work in vector components in nonplanar systems.

$$F_{net} = \Sigma F = (\Sigma F_x)i + (\Sigma F_y)j + (\Sigma F_z)k = 0$$

F_x, F_y, and F_z represent the force scalar components in the x-, y-, and z-directions.

$$\Sigma F_x = 0 \quad \Sigma F_y = 0 \quad \Sigma F_2 = 0$$

Moment Equilibrium Equations (for Moment Balance)

$$M_{net} = \Sigma M = (\Sigma M_x)i + (\Sigma M_y)j + (\Sigma M_z)k = 0$$

M_x, M_y, and M_z represent the individual moment components about the x, y, and z axes at a specific reference point.

$$\Sigma M_x = 0 \quad \Sigma M_y = 0 \quad \Sigma M_z = 0$$

SUMMARY

The necessary and sufficient condition for the static equilibrium of a body are that the resultant force and couple from all external forces form a system equivalent to zero:

$$\sum \vec{F} = 0 \sum \vec{M}_O = \sum (\vec{r} \times \vec{F}) = 0$$

Resolving each force and moment into its components results in six scalar equations

$$\Sigma F_x = 0 \quad \Sigma F_y = 0 \quad \Sigma F_z = 0$$
$$\Sigma M_x = 0 \quad \Sigma M_y = 0 \quad \Sigma M_z = 0$$

SECTION 4 Problems

4.1. SIMPLY SUPPORTED BEAM

Problem Author: Niara Comrie

Problem Statement:
Stacy is a gymnastic who is performing a flip on her beam. She applies a downward force of 500 N on the beam. Her force causes a moment at B. The beam is supported by a pin and a roller. What are the unknown forces?

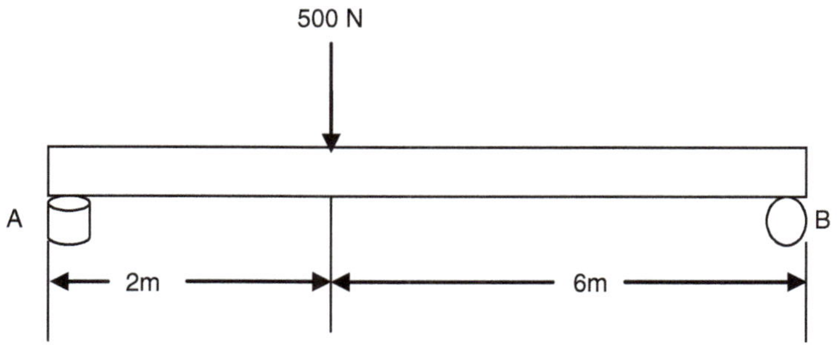

Goal:
To solve A_y, A_x, F_A, F_B, and B_y

Given:
500 N, Roller and Pin

Assumptions:
Assuming that the gymnast is only applying one force.

Equations Needed:
Sum of forces in x- and y-directions and the moment equation

Solution:

FBD

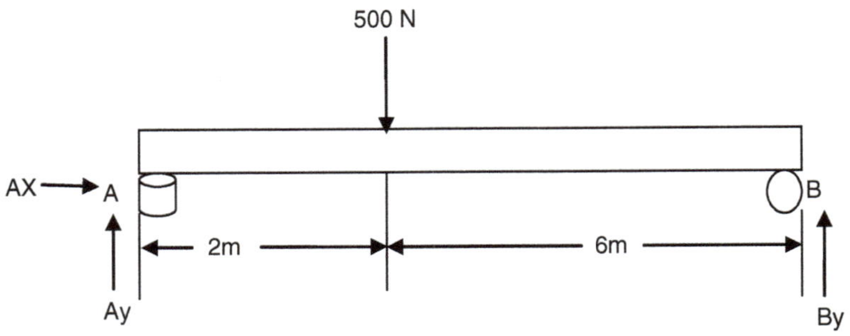

$$F_x = 0$$

$$A_x = 0$$

$$F_y = 0,\ A_y + B_y - 500 = 0$$

$$M @ B = -500(2) + A_y(6) = 0$$

$$6(A_y) = 1{,}000$$

$$A_y = \frac{1{,}000}{6} = 166.66 \text{ N}$$

$$166.66 - 500 = -B_y$$

$$B_y = 333.3 \text{ N}$$

$$F_A = \sqrt{A_x^{\ 2} + A_y^{\ 2}} = \sqrt{0^2 + 166.66^2} = 0$$

$$= 166.66 \text{ N}$$

$$F_B = \sqrt{B_x^{\ 2} + B_y^{\ 2}} = \sqrt{0^2 + 333.33^2} = 0$$

$$= 333.33 \text{ N}$$

4.2. EQUILIBRIUM OF A BOAT

Problem Author: Richard Blorstad

Type of Problem:
FBD with a distributed load

Problem Statement:
In rowing (crew), it is very important that the boat and all its participants be in equilibrium. Draw the FBD of the boat, assuming all of its participants weigh 800 N, and the boat is 15-m long and 0.33-m wide. Assuming the boat can be approximated as a rectangular prism (or cuboid), how much does it weigh if the boat and all its participants sits 10 cm in the water? You may assume the buoyancy force is constant across the entire length of the boat and is evenly distributed.

Drawing:

Goal:
Solve for the weight of the boat (unknown)

Given:
Five forces (due to the rowers) and an evenly distributed buoyancy force (found using the equation for buoyancy force).

Assumptions:
Boat can be approximated as a rectangular prism

Equations Needed:

$$F_{\text{buoyancy}} = \rho_{H_2O} \times g \times V_{\text{displaced fluid}}$$

Answer:

$$F_{\text{buoyancy}} = \left(1{,}000\ \frac{\text{kg}}{\text{m}^3}\right) \times \left(9.81\frac{\text{m}}{\text{s}^2}\right) \times (15 \times 0.33 \times 0.1\,\text{m}^3) = 4{,}851\,\text{N}$$

$$F_{\text{buoyancy distributed}} = \frac{4{,}851\,\text{N}}{15\,\text{m}} = 323.4\ \frac{\text{N}}{\text{m}}$$

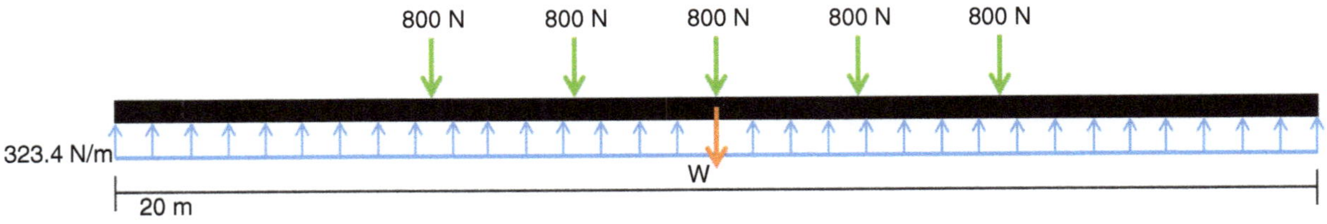

$$\Sigma F_y = 0 = 4{,}851\,\text{N} - 5 \times (800\,\text{N}) - \text{W} \qquad W = 851\,\text{N}$$

4.3. FREE-BODY DIAGRAMS OF MULTIPLE SYSTEMS

Problem Author: Jordyn Schroeder

Type of Problem:

2D moments
Students should be able to understand FBDs

Problem Statement:

For the system shown in the following, the dumbbell at *C* has a weight of 95 lb. and the pulley has a weight of 10 lb. Draw the FBDs for:

1) Ring at *B*
2) Cable *BC*
3) Dumbbell at *C*
4) Pulley at *D*

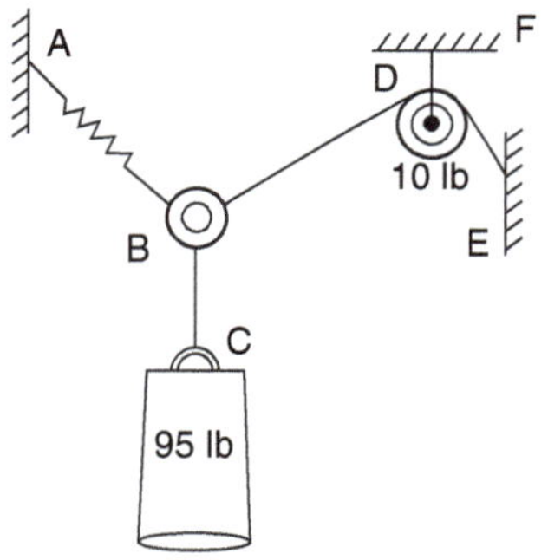

Goal:

Draw requested FBDs

Given:

F_{grav} of Dumbbell = 95 lb., F_{grav} of pulley = 10 lb.

Assumptions:

N/A

Equations Needed:

N/A

Answer:

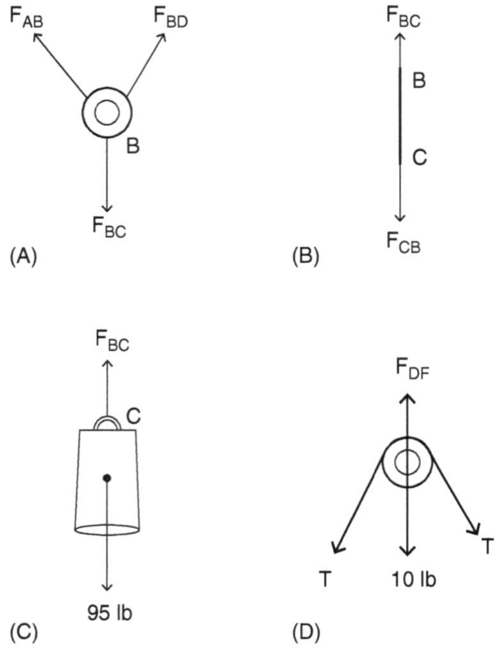

4.4. ANALYSIS OF A ROLLER COASTER

Problem Author: Robert Oehrli

Type of Problem:
This problem gives students practice solving FBDs and solving systems of equations.

Problem Statement:
The rollercoaster train is carried up the lift hill a chain at 40°. The empty rollercoaster train weighs 2,500 kg and can hold 10 passengers weighing 70 kg each. Part I: Draw an FBD that shows all of the forces acting on the train. Part 2: Solve for all of the forces acting on the train.

Drawing:

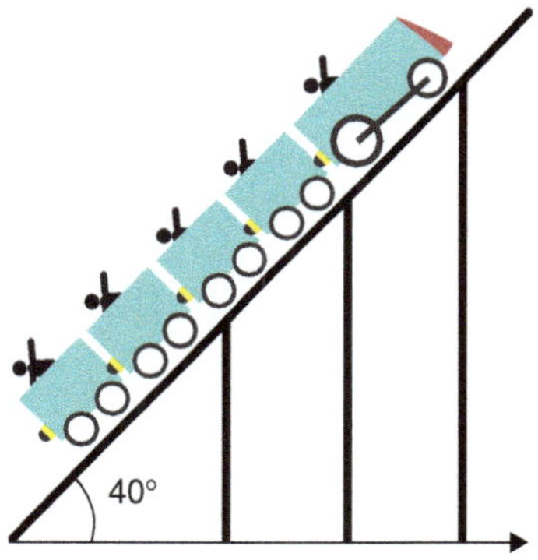

Given:
Passenger Weight: 70 kg
Number of Passengers: 10
Weight of Empty Train: 2,500 kg
$\mu_s = 0.55$

Assumptions:
Assume there is no drag and that the friction is static friction.

Equations Needed:

$$\sum F_x = 0$$

$$\sum F_y = 0$$

$$F_{\text{Friction}} = \mu \times F_{\text{Normal}}$$

Answer:

Step 1: Draw the FBD.

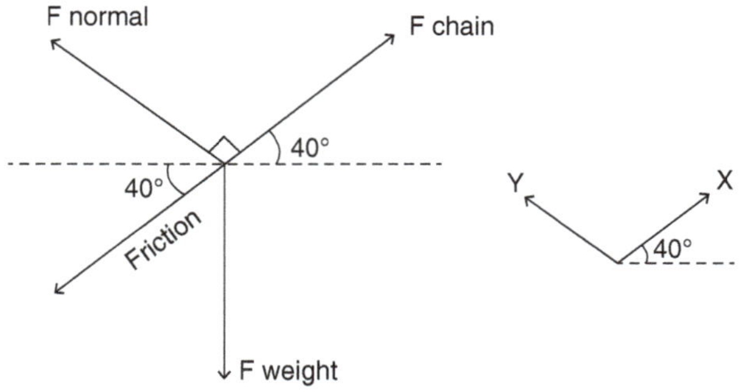

In order to minimize the amount of trig involved with this problem, we can simply define our x-axis and y-axis to be tilted at 40°.

Step 2: Write out the system of equations.

Equation 1: $\sum F_x = 0 = F_{\text{Chain}} - F_{\text{Friction}} - \text{Weight} \times \cos(50°)$

Equation 2: $\sum F_y = 0 = F_{\text{Normal}} - \text{Weight} \times \sin(50°)$

Equation 3: $F_{\text{Friction}} = \mu \times F_{\text{Normal}}$

Equation 4: $\text{Weight} = \text{Mass} \times g$

Step 3: Solve for weight using Equation 4.

$$\text{Mass} = \text{Mass of Train} + \text{Mass of Pasengers}$$

$$\text{Mass} = 2{,}500\,\text{kg} + (10)70\,\text{kg}$$

$$\text{Mass} = 3{,}200\,\text{kg}$$

$$\text{Weight} = 3{,}200\,\text{kg} \times 9.8\,\frac{\text{m}}{\text{s}^2}$$

$$\text{Weight} = 31.36\,\text{kN}$$

Step 4: Solve for the normal force using Equation 2.

$$\sum F_y = 0 = F_{\text{Normal}} - \text{Weight} \times \sin(50°)$$

$$F_{\text{Normal}} = \text{Weight} \times \sin(50°)$$

$$F_{\text{Normal}} = 31.36\,\text{kN} \times \sin(50°)$$

$$F_{\text{Normal}} = 24.02\,\text{kN}$$

Step 5: Solve for the friction force using Equation 3.

$$\sum F_y = 0 = F_{\text{Normal}} - \text{Weight} \times \sin(50°)$$

$$F_{\text{Normal}} = \text{Weight} \times \sin(50°)$$

$$F_{\text{Normal}} = 31.36\,\text{kN} \times \sin(50°)$$

$$F_{\text{Normal}} = 24.02\,\text{kN}$$

Step 6: Solve for the friction force using Equation 1.

$$\sum F_x = 0 = F_{\text{Chain}} - F_{\text{Friction}} - \text{Weight} \times \cos(50°)$$

$$F_{\text{Chain}} = F_{\text{Friction}} + \text{Weight} \times \cos(50°)$$

$$F_{\text{Chain}} = 13.21\,\text{kN} + 31.36\,\text{kN} \times \cos(50°)$$

$$F_{\text{Chain}} = 31.36\,\text{kN}$$

4.5. ANALYSIS INVOLVED IN BIOMECHANICS

Problem Author: Prof. Jamie Gurganus

Type of Problem:
2D rigid-body planar problem

Problem Statement:
An arm is holding a 1-kg weight using muscles within the forearm. The bicep, which connects the forearm to the upper arm, pulls with a tension force B located 2.75 cm away from the elbow. Calculate the muscle force B, needed to maintain the forearm in the 15° position. The length from the elbow to the center of the 1-kg weight is 30.5 cm. Weight of the arm is 15.7 N.

Assumptions:
The elbow is a pin. The hand and the forearm are a rigid-body planar system.
B = Bicep

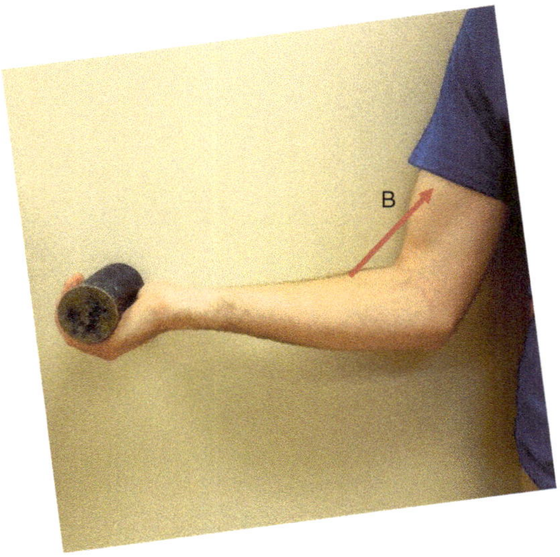

Courtesy of Jamie Gurganus

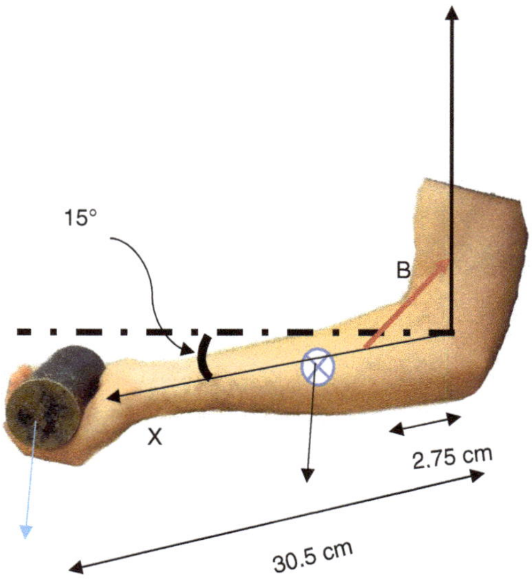

4.6. 3D ANALYSIS OF A TRAP DOOR

Problem Author: Prof. Gurganus

Type of Problem:
3D rigid-body equilibrium problem

A 100 lb. trap door is opened and held by lock hinges A and B. Find the reaction forces of the hinges at A and B. B only resist forces in the x- and y-directions while A resist in the x-, y-, and z-directions.

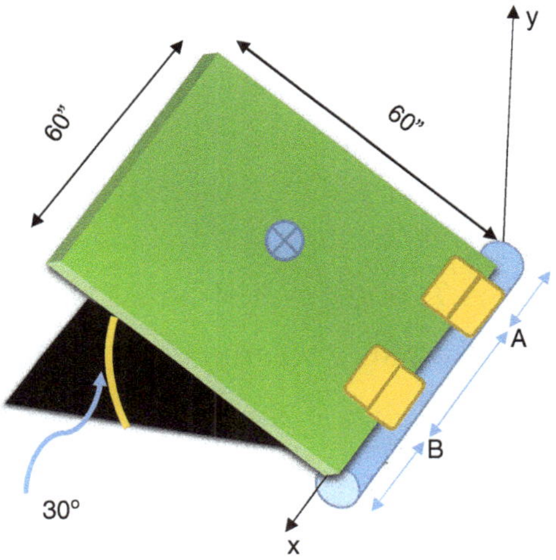

4.7. 3D ANALYSIS ON A DRIVE SHAFT

Problem Author: Dr. Neil S. Rothman

Type of Problem:

This problem covers the concept of creating an FBD for a circular shaft subjected to multiple loads in different orientations.

The student should be able to use summation of forces and moments (torques) to find the unknown loads on the second pulley and the reaction forces at the bearings. These should then be correctly shown on an FBD as concentrated moments at the pulleys, net forces at the pulleys, and reaction forces at the bearings in two planes.

Problem Statement:

The following is a shaft supporting two pulleys with two belts. The forces shown are produced by the belts and are in equilibrium. Tension $T_2 = 0.15\ T_1$.

A) Determine what the tensions T_1 and T_2 must be for the pulleys to be in equilibrium.

B) Draw the FBD for the shaft. You may assume that the bearings at the ends provide simple supports for the shaft.

Drawing:

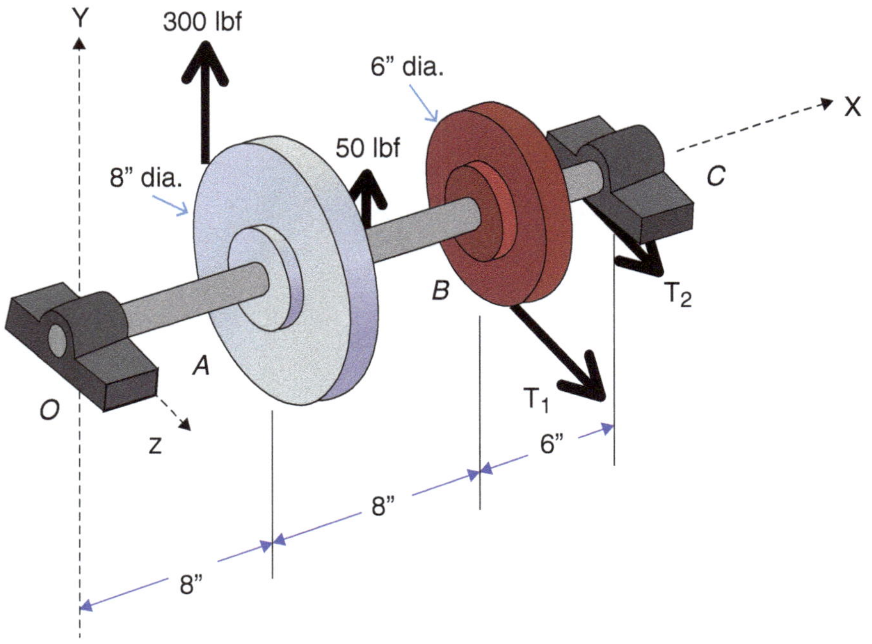

Goal:

1) Determine the magnitudes of T_1 and T_2 to achieve equilibrium with the other applied forces.
2) Draw an FBD of the shaft by calculating the reaction forces at the bearings at O and C.

Given:

1) Dimensions of shaft and pulleys
2) Forces applied to the pulleys with $T_2 = 0.15\ T_1$

Assumptions:

There are no applied forces in the X-direction, so there are no reaction forces in the X-direction at the bearings at O and C.

The bearings provide simple supports (i.e., no moment reactions)

Equations Needed:

$\sum T = 0$ (summation of torques about the X-axis)

$$\sum F_Y = 0$$

$$\sum F_Z = 0$$

$$\sum M_Y = 0$$

$$\sum M_Z = 0$$

Answer Part A:

For the pulleys to be in equilibrium, the total of the torques applied to the shaft must equal zero.

$\sum T\ =\ 0$ For pulley A, the radius is 4 in. and there are two forces acting tangent to the pulley and parallel to each other. Hence, the torque applied to pulley A is: $T_A =\ (300\,\text{lbf})(4\,\text{in.}) - (50\,\text{lbf})(4\,\text{in.}) = 1{,}000\,\text{in.} - \text{lbf}$ (adhering to the positive direction per right-hand rule)

For equilibrium, $T_A + T_B = 0$, or $T_B = -1{,}000$ in-lbf

Similarly, since the forces applied to pulley B are tangent and parallel

$$T_B =\ (T_2)(3\,\text{in.}) - (T_1)(3\,\text{in.})$$

Substituting in for $T_2 = 0.15\ T_1$ and setting it equal to T_A

$$T_B = (0.15\ T_1)(3\,\text{in.}) - (T_1)(3\,\text{in.}) = -0.85\ T_1 = -1{,}000\,\text{in-lbf}$$

$\rightarrow T_1 = 1{,}176.5$ lbf and therefore $T_2 = (0.15)(1{,}176.5) = 176.5$ lbf Ans.

Part B:

Draw the shaft as a line and add the forces and torques being applied to it. The bearings at O and C should be replaced with reaction forces in the Y- and Z-directions since there are applied loads in those directions. This should produce a figure as shown in the following. Note that the load in the Y-direction at pulley A is the sum of the two loads applied to the pulley; similarly at pulley B.

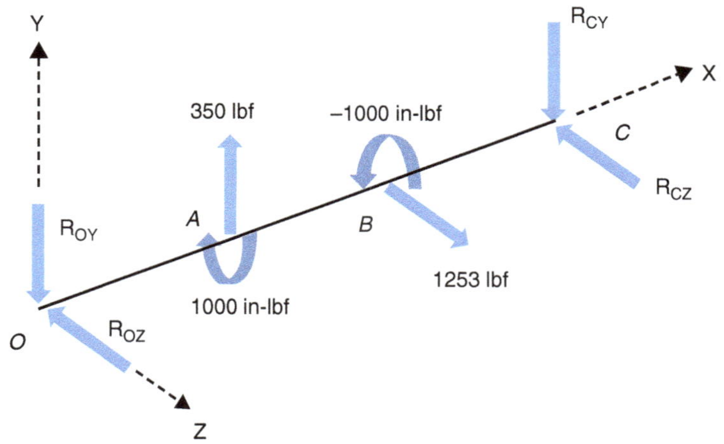

To solve for the reaction forces at O and C, we must use summation of forces in the Y- and Z-directions and summation of moments about the Y and Z axes centered at the origin.

$$\sum F_Y = 0 = -R_{OY} + 350 - R_{CY}$$

$$\sum F_Z = 0 = -R_{OZ} + 1{,}253 - R_{CZ}$$

$$\sum M_Y = 0 = (350)(8) - (R_{CY})(22) \rightarrow R_{CY} = 127.3\,\text{lbf}$$

$$\sum M_Z = 0 = (1{,}253)(16) - (R_{CZ})(22) \rightarrow R_{CZ} = 911.3\,\text{lbf}$$

Substituting back in the force balance equations gives the other reaction forces:

$$0 = -R_{OY} + 350 - 127.3 \rightarrow R_{OY} = 222.7\,\text{lbf}$$

$$0 = -R_{OZ} + 1{,}253 - 911.3 \rightarrow R_{OZ} = 341.7\,\text{lbf}$$

Note that the signs for each of the reaction forces being positive indicate that the assumed directions are correct.

Summation of the reaction forces in the Y-direction and Z-direction gives confirmation that equilibrium is achieved.

Ans.
This completes the FBD and the problem.

Chapter 5

Frames, Machines, and Trusses

FRAMES

Frames are Systems That are Designed to Support Any Loads

Unlike a single member system, frames are made up of several members (parts) that hold it together. Frames can be anything from a shelf on the wall to a deck on the house. The members of the frame exert forces on each other that are equal and opposite (recall Newton's 3rd Law).

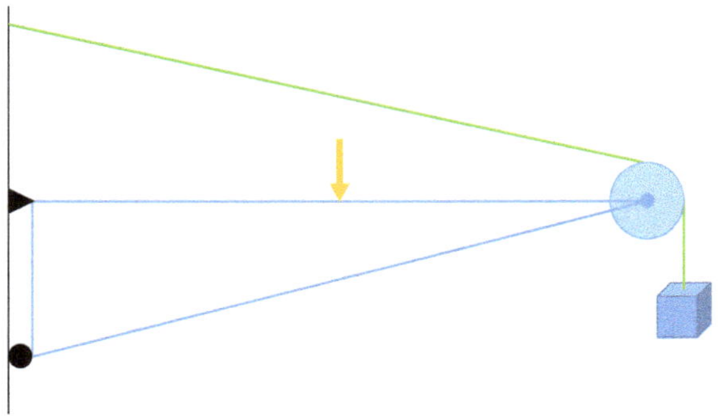

MACHINES

Machines are Used to Manipulate Loads on the System

In a machine, an input load is applied *into* the system and an output load is produced *out* of the system. Machines are often used to maximize the ratio of the output to the input.

Think of a lever (as shown on the following page). The farther away your hand is from the axis of rotation (distance or length), the greater the moment is produced. (recall that moments increase as the force moves farther away from the moment center). Hence why levers are considered simple machines!

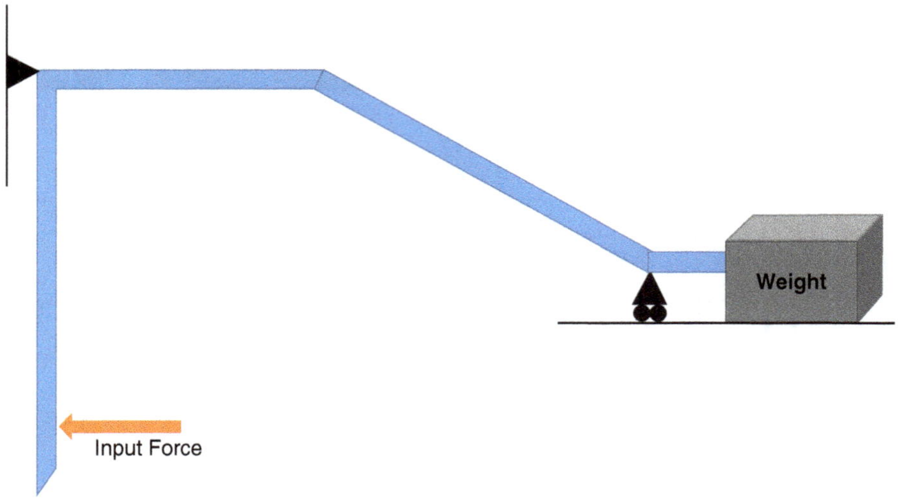

A common simple machine utilized in frames and machines is the pulley (as shown in the following).

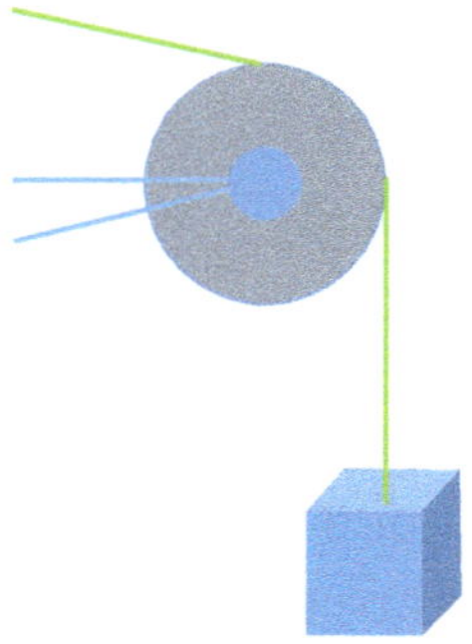

Pulleys are used to transfer a load into a new direction and when used with other pulleys, can be used to lift heavy objects with ease. In statics, you likely find that a Pulley is considered to be frictionless and massless.

Note: Pulley's can change direction of the tension, however, the magnitude will remain the same (ONLY with massless frictionless pulleys). Remember tension is a force (vector).

An example of a pulley is shown in the following frame example.

Similarities and Differences Between the Frame and Machines

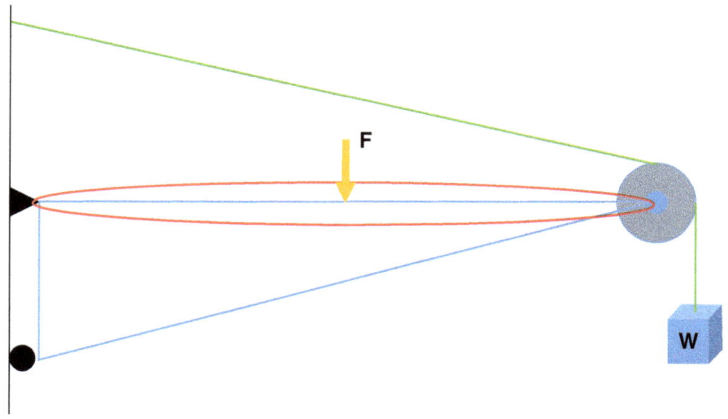

To be classified as a frame, at least *one* member must be a multiforce member. Multiforce means that the member *must have* more than two forces acting on it. The forces can be from other members of the frame or external loads.

The preceding frame example has a single multiforce member, with forces from the pin, the pulley, and the external load acting on the circled member. The forces within the members change with the external loads on the system. As the external loads increase, so must the forces within the system, in order to support the loading.

Just like with frames, machines also are made up of multiple members, one of which must be a multiforce member. Machines, however, are meant to transfer loads, instead of support them. Can we classify the figure on the previous page as a machine? Why?

METHODS FOR DETERMINING LOADS IN FRAME AND MACHINE (BOTH PLANAR AND NONPLANAR SYSTEMS)

Frames

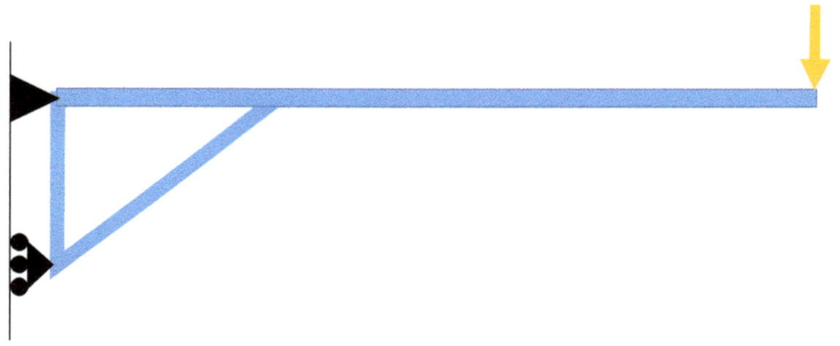

Methods for determining internal loads in a frame:

- Identify which members (if any) are zero-force or two-force members. This will greatly simplify the problem later!
- Next, identify the supports in the system and the types of support loads (forces and moments).
- Draw a free body diagram (FBD) for each multiforce member of the frame. Be careful to use equal and opposite forces when drawing the internal forces (green) between members (figures shown). Additionally, notice the external forces (yellow).

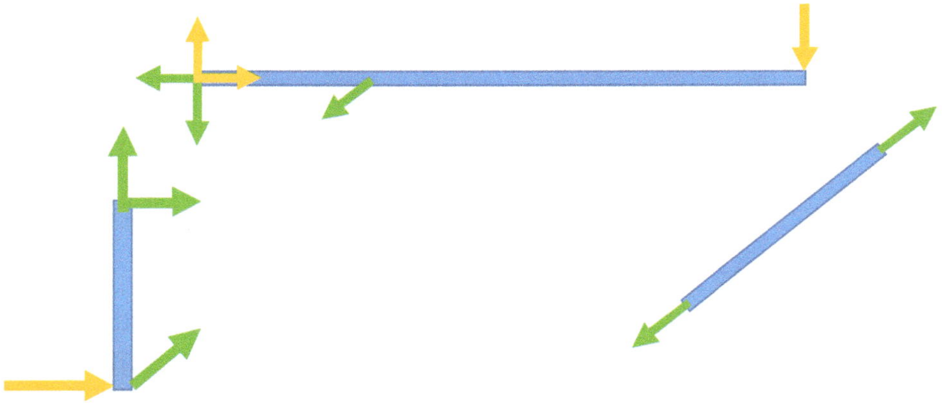

- Set up your equations of equilibrium (*E* of *E*'s) and a strategy to solve them. Solve the equation that has only one unknown, then use that value to solve the remaining equations.

Machines

Analyzing the internal forces in a machine is exactly the same as in a frame. Unlike in frames, however, you will often have to solve for what the input force or output force is in order for a certain condition to be met.

TRUSS

A truss is a system made up of many two-force members (Truss A and B) connected by pins. They can be supported by pins, which restrict linear movement in the x and y planes, or by a roller which restricts movement perpendicular to the surface on which it rests.

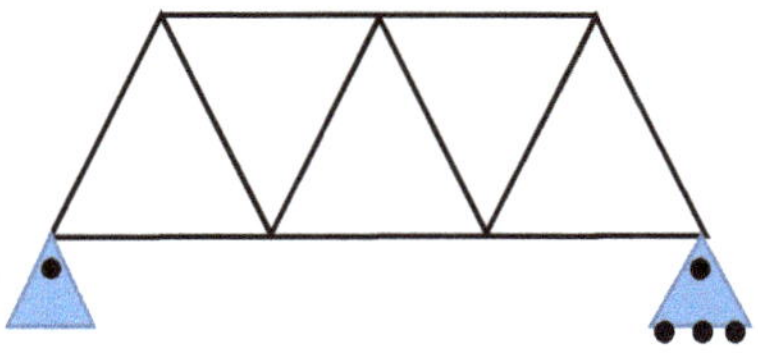

Truss A

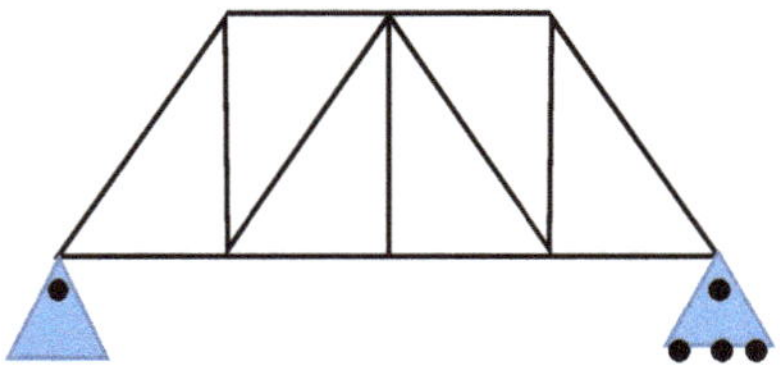

Truss B

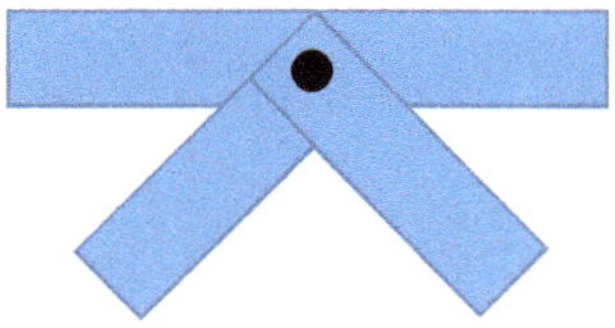

Members connected by a pin

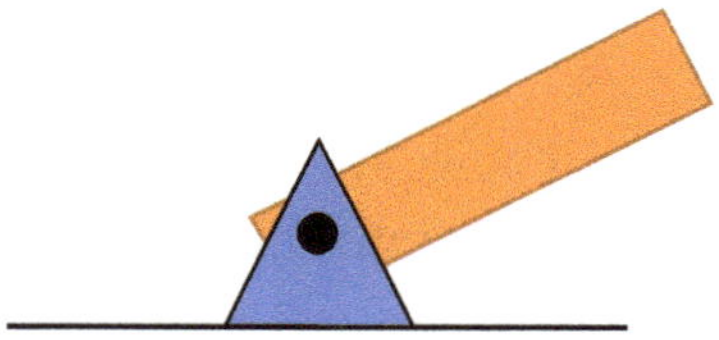

Pin

Roller

In a truss system, all of the loads MUST act at a pin. In comparing to a frame, the loads act along the member.

Internal Forces in Trusses

Internal forces on a truss are created when a force on each end of the member either pull or push in opposite directions. This is indicated by the green arrows as shown in the following.

Tension: If the force points away from the pin toward the center of the member, then that member will produce an opposing force.

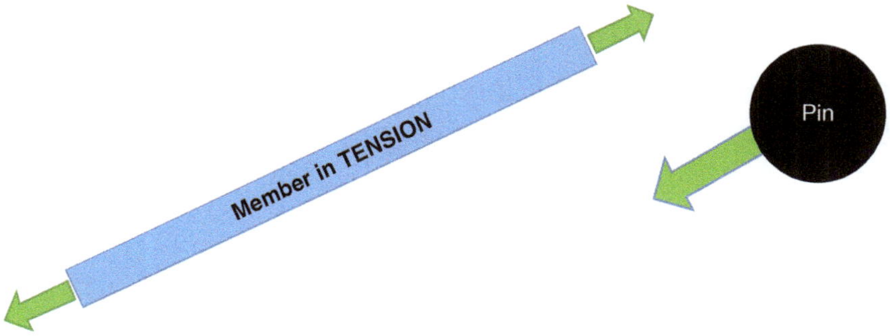

An example is the game of tug of war you have two opposite sides pulling on a rope. This pull creates a tension force.

However, if the force (arrow) acts toward the pins and away from the center of the member then that member is in compression as illustrated. For example, when crushing an aluminum soda can, there are two opposite forces acting on the can causing it to crush under a compressive force.

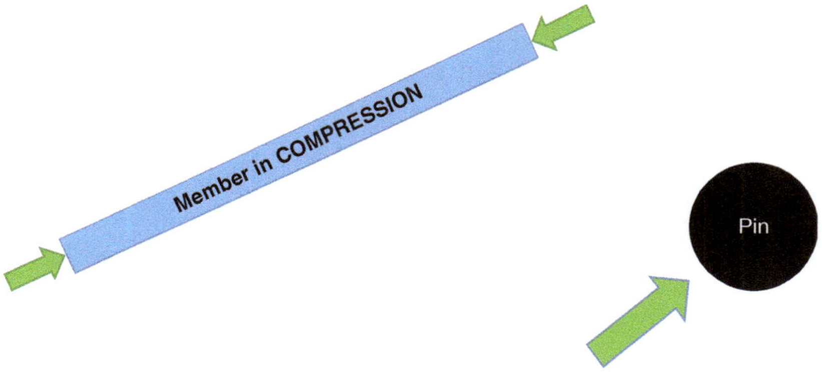

EVALUATION AND ASSESSMENT OF TRUSSES

The Method of Joints

The Method of Joints is typically used in a situation where the internal forces for all of the members of a truss are needed.

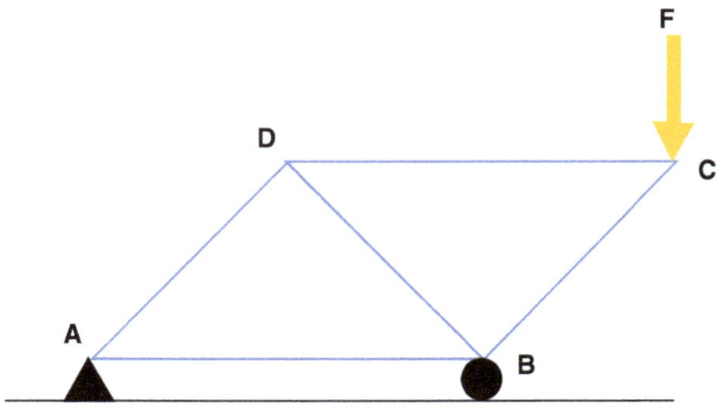

Steps for Analysis for Method of Joints

There are multiple ways to determine the internal loads on each member using the method of joints. Two are outlined here. *Note:* When using large systems of equations, which can happen in this method of assessment, it is often helpful to number each equation making it easier to keep track of the joints.

The first method takes you through the process one joint at a time.

1. Draw an FBD for the entire truss including only the external forces acting on the truss.

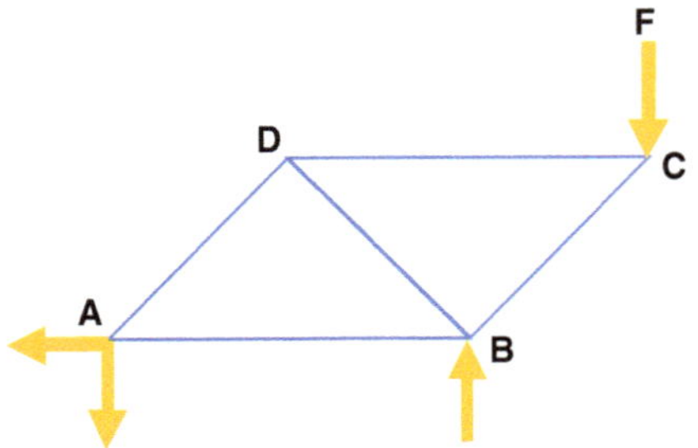

2. Solve for the reaction forces at each support using the Equations of Equilibrium.
3. Next, draw an FBD for the simplest joint, one that has only one or two unknowns.

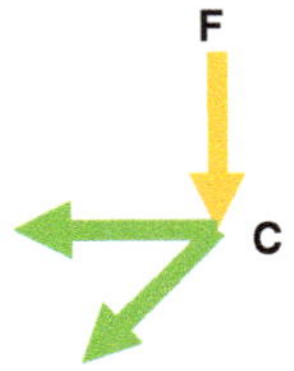

   Helpful hint: Draw all of your forces in the members in tension. This will make problem solving easier. Remember, a forces' magnitude is never negative. A positive result is classified as tension where a negative answer is in compression.
4. Use the equations of equilibrium to solve the internal forces in the X and Y directions.
5. Continue to solve the remaining joints in the truss system using steps 3 and 4.

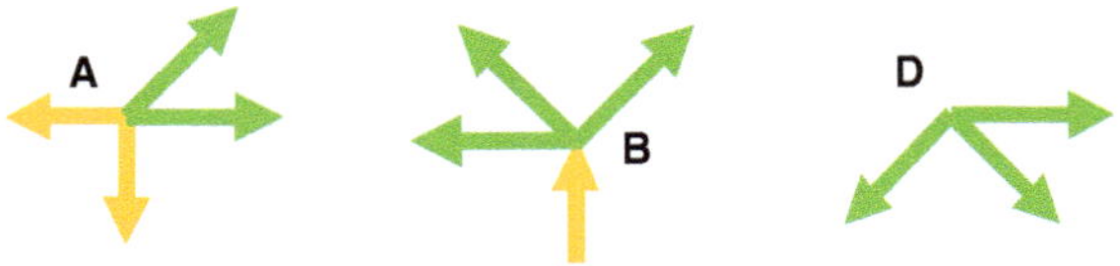

6. Solve any remaining equations.

The second method of joints solves the entire system in less steps.

1. Draw an FBD for the entire truss including only the external forces acting on the truss.
2. Solve for the reaction forces at each support.

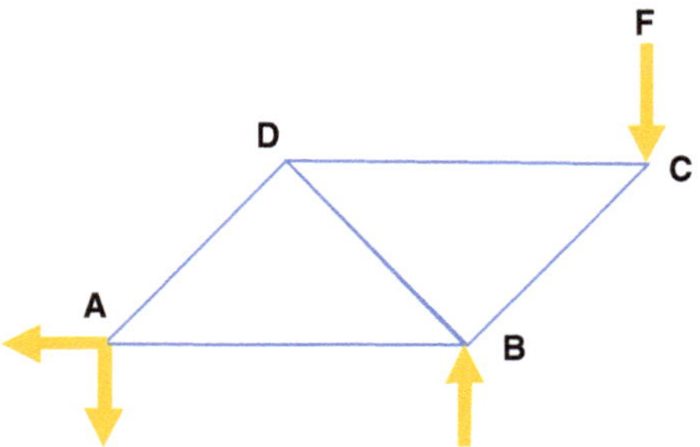

3. Draw an FBD for all of the joints.

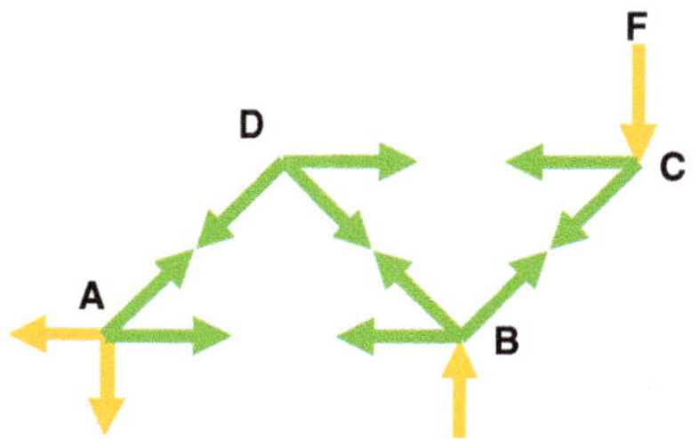

Draw all of your members as if they are in tension.
Determine the equations of equilibrium in the X and Y directions for all of the joints and make sure to number each equation.
4. Solve the system of equations. Remember: If you get a positive answer, the force is in tension otherwise, if it results in a negative answer, it's in compression. This is assuming you drew all your forces in tension.

The Method of Sections

Method of Sections is typically used when you need to determine specific internal loads in a truss.

Steps for Analysis for Method of Sections

1. Draw an FBD for the entire truss including only the external forces acting on the truss.

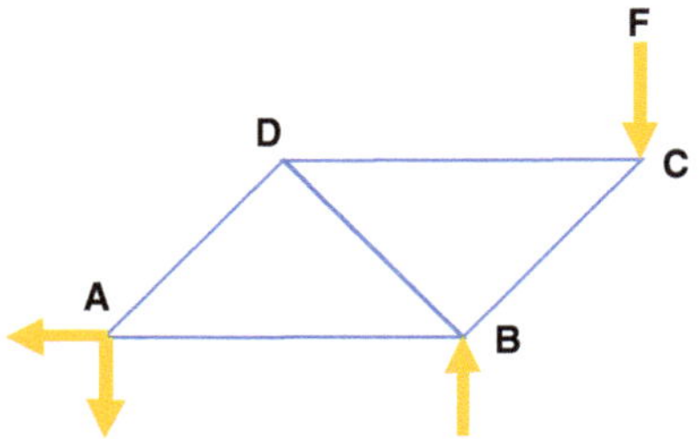

2. Solve for the reaction forces at each support.
3. Decide which members of the truss to draw your cut line along. The cut line should intersect three members. Recall: 3 equations, 3 unknowns.

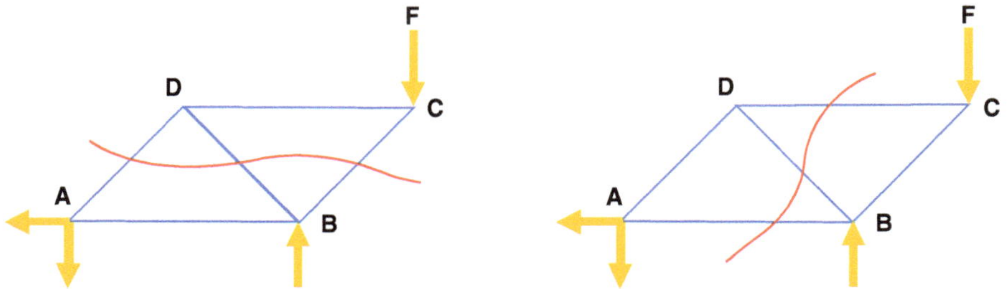

4. Draw an FBD for each side of the truss. Include the internal forces along the cut line.

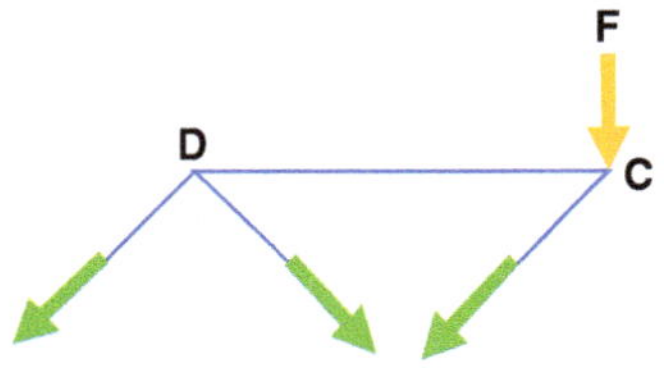

5. Pick the side of the truss to analyze that has the fewest internal loads and support loads.

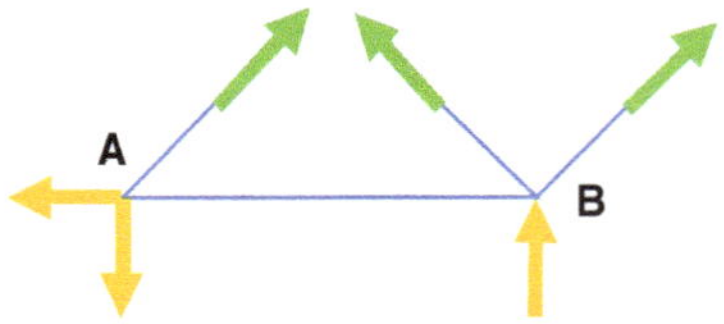

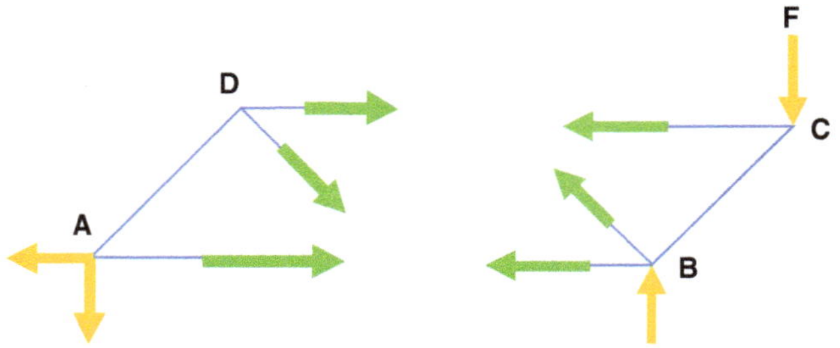

6. Solve for all of the unknown forces using the equations of equilibrium.

Zero Force Members

A zero force member is neither in compression or tension (no-load). In statics you will come across the following two cases:

A. When two members meet at a joint that has no external load or support load at that joint, then both members are zero force members.
B. When three members meet at a joint that has no external load or support load acting upon and two of the members are colinear, than then the third member is a zero force member.

If you determine there are zero force members in a truss then you should remove them and simplify the truss before you begin static analysis.

Section 5 Problems

5.1. FRAMES WITH TWO-FORCE MEMBERS

Problem Author: Jordyn Schroeder

Type of Problem:
Students should be able to understand FBDs, frames, and two force members.

Problem Statements:

1) Shown in the following, a stick figure stands at the end of a rigid diving board. If the supporting beam *BC* can withstand 1,000 N of force before breaking, what is the maximum force that can be applied by the stick figure.
2) A 50 kg stick figure is standing at the end of a rigid diving board. Solve for the reaction forces at *A* and force on member *BC*

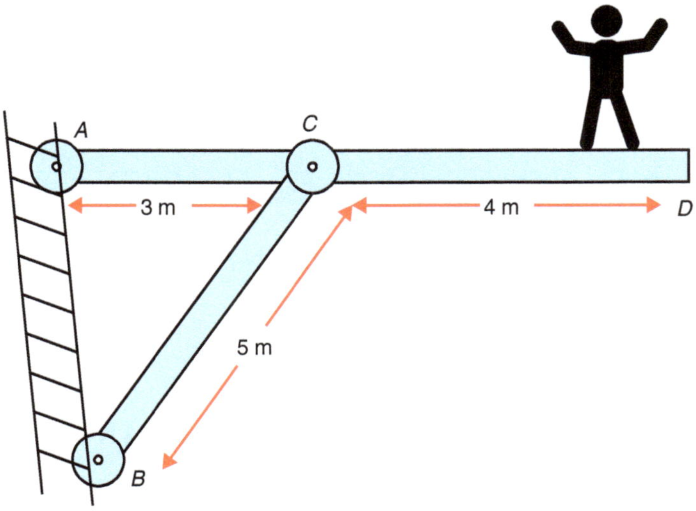

Goal:
(1) assuming $\boldsymbol{F}_{BC}$ = 1,000 N, find F_{SF} (2) find $\boldsymbol{F}_{BC}$, A_X, and A_Y

Given:
(1) lengths of all members and $\boldsymbol{F}_{BC}$=1,000 N (2) lengths of all members and mass of stick figure = 50 kg

Assumptions:
Member BC is a two force member

Equations Needed:
$\Sigma M = 0,\ \Sigma \boldsymbol{F}_x = 0,\ \Sigma \boldsymbol{F}_y = 0$

Answer:
The FBDs of each member is shown in the following: M_A

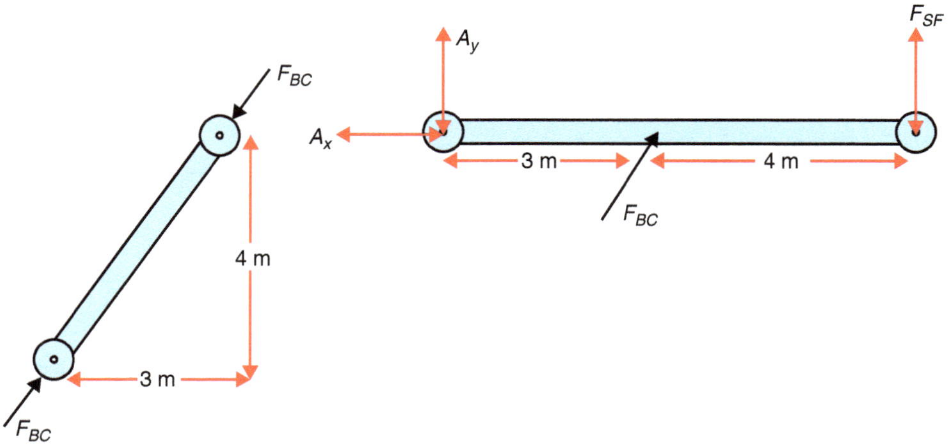

1) $\Sigma M_A = 0 = \left(\frac{4}{5}\right)(\boldsymbol{F}_{BC})(3\,\text{m}) - (\boldsymbol{F}_{SF})(7\,\text{m}) = \left(\frac{4}{5}\right)(1{,}000\,\text{N})(3\,\text{m}) - (\boldsymbol{F}_{SF})(7\,\text{m}) \rightarrow \boldsymbol{F}_{SF} = 342.9\,\text{N}$

2) $\Sigma M_A = 0 = \left(\frac{4}{5}\right)(\boldsymbol{F}_{BC})(3m) - (\boldsymbol{F}_{SF})(7\,\text{m}) = \left(\frac{4}{5}\right)(\boldsymbol{F}_{BC})(3\,\text{m}) - (50\,\text{kg})\left(\frac{9.8\,\text{m}}{s^2}\right)(7\,\text{m})$

$$\rightarrow \boldsymbol{F}_{BC} = 1{,}429.17\,\text{N}$$

$$\Sigma \boldsymbol{F}_x = 0 = -A_x + \left(\frac{3}{5}\right)(\boldsymbol{F}_{BC}) = -A_x + \left(\frac{3}{5}\right)(1{,}429.17\,\text{N}) \rightarrow A_x = 857.5\,\text{N}$$

$$\Sigma \boldsymbol{F}_y = 0 = A_y + \left(\frac{4}{5}\right)(\boldsymbol{F}_{BC}) - \boldsymbol{F}_{SF} = A_y + \left(\frac{4}{5}\right)(1{,}429.17\,\text{N}) - (50\,\text{kg})\left(\frac{9.8\,\text{m}}{s^2}\right)$$

$$\rightarrow A_y = -653.34\,\text{N} = 653.34\,\text{N}$$

5.2. FAILURE (OR NOT) WITHIN A FRAME

Problem Author: Sarah Hemler

Type of Problem:
The student should be able to calculate the system's reaction forces.

Problem Statement:
The outdoor lamp is held to the bar at point A via a pin. The bar is fixed to the wall with pins at B and C. The outdoor lamp weighs 10 lb. Assume the vertical reaction component of pin C is at its maximum of 23 lb. (a) Determine the reaction forces at B and C. (b) If a 27 lb electric fly zapper is hung from the lamp and at B, the maximum horizontal reaction force is 50 lb, will the mechanism fail?

Drawing:

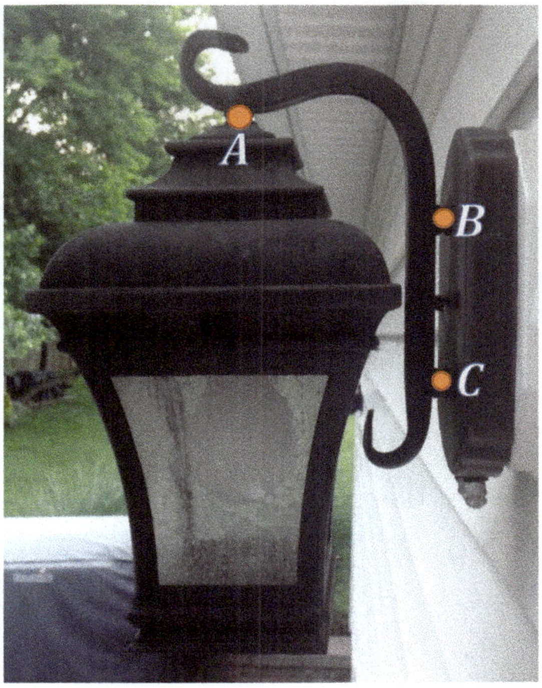

Courtesy of Sarah Hemler

Solution:
a) $B_x = -13.33\,\text{lb.}, B_y = -13\,\text{lb.}, C_x = 13.33\,\text{lb.}, C_y = 23\,\text{lb.}$
b) $B_x = -50.67\,\text{lb.},$

Goal:

Find the reaction forces at pin B and C

Given:

(a) $C_y = 23\,\text{lb.}$, (b) $|B_{x_{\max}}| = 50\,\text{lb.}$

Assumptions:

No friction within the pins

Equations Needed:

$\sum \boldsymbol{F}_x = 0,\ \sum \boldsymbol{F}_y = 0,\ \sum M_z = 0$

Answer:

Draw the FBD:

Courtesy of Sarah Hemler

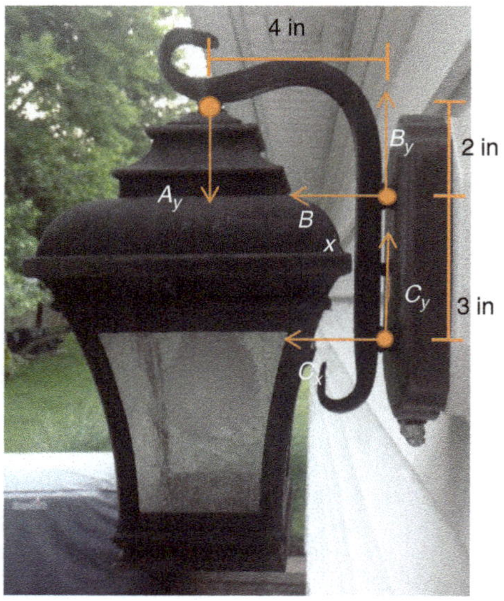

Courtesy of Sarah Hemler

a) At pin A (left image):

$$\sum \boldsymbol{F}_y = 0 = A_y - 10\,\text{lb.} \rightarrow A_y = 10\,\text{lb.}$$

Equations of equilibrium for system (right image):

1) $\sum \boldsymbol{F}_x = 0 = -B_x - C_x \quad \rightarrow B_x = -C_x$

2) $\sum M_B = 0 = -3\,\text{in.} \times C_x - 4\,\text{in.} \times A_y \quad \rightarrow C_x = 13.33\,\text{lb.} \rightarrow \text{from (1) } B_x = -13.33\,\text{lb.}$

3) $\sum \boldsymbol{F}_y = 0 = B_y + C_y - A_y \quad \rightarrow B_y = A_y - C_y = 10\,\text{lb.} - 23\,\text{lb.} = -13\,\text{lb.}$

b) At pin A:

$$\sum \boldsymbol{F}_y = 0 = A_y - 10\,\text{lb.} - 27\,\text{lb.} \rightarrow A_y = 37\,\text{lb.}$$

Equations of equilibrium for system (right image):

1) $\sum \boldsymbol{F}_x = 0 = -B_x - C_x \quad \rightarrow B_x = -C_x$

2) $\sum M_B = 0 = -3\,\text{in.} \times C_x - 4\,\text{in.} \times A_y \quad \rightarrow C_x = 50.67\,\text{lb.} \rightarrow \text{from (1) } B_x = -50.67\,\text{lb.}$

Since the horizontal reaction force at *B* is greater than 50 lb., the mechanism will fail.

5.3. REACTION FORCES WITHIN A TEAPOT

Problem Author: Sarah Hemler

Type of Problem:

The mechanism that is used to open the spout of a teapot can be classified as a machine. The student should be able to calculate the system's internal reaction forces.

Problem Statement:

In order to keep the spout cover of a teapot open, a force of 10 N must be exerted 30° from the horizontal surface on the handle. Find the reaction forces at each of the pins.

Drawing:

Courtesy of Sarah Hemler

Courtesy of Sarah Hemler

Solution:

$A_x = B_x = 8.66\,\text{N}$, $A_y = B_y = 5.00\,\text{N}$

Goal:

Find the reaction forces at pin A and B

Given:

Pulling force, P and angle of elevation, θ

Assumptions:

No friction within the pins

Equations Needed:

$\sum F_x = 0, \sum F_y = 0, \sum M_z = 0$

Answer:

Separate the machine parts and create the appropriate FBDs

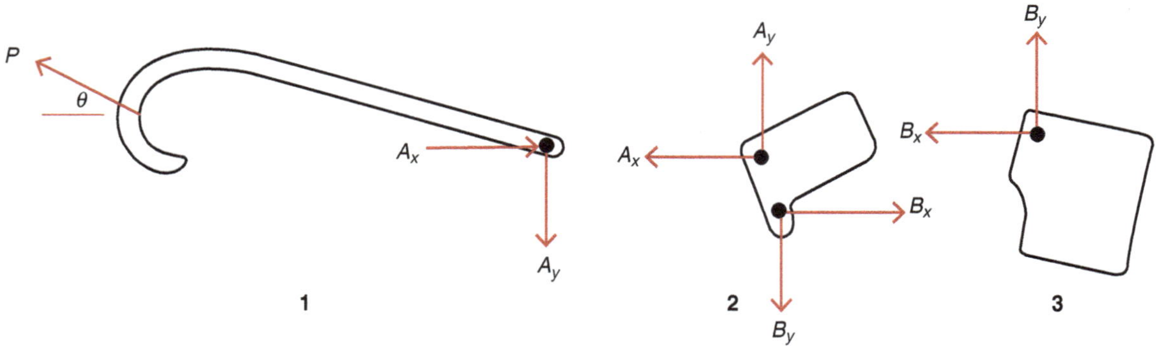

Analyze FBD 1:

$$\sum F_x = 0 = A_x - P\cos(\theta) \rightarrow A_x = 10\,\text{N} \times \cos(30°) = 8.66\,\text{N}$$

$$\sum F_y = 0 = -A_y + P\sin(\theta) \rightarrow A_y = 10\,\text{N} \times \sin(30°) = 5.00\,\text{N}$$

Analyze FBD 2:

$$\sum F_x = 0 = B_x - A_x \rightarrow B_x = 8.66\,\text{N}$$

$$\sum F_y = 0 = A_y - B_y \rightarrow B_y = 5.00\,\text{N}$$

5.4. FORCES WITHIN A TRUSS

Problem Author: Nicole Demetrides

Type of Problem:

This problem covers trusses. The student should be able to utilize method of sections and joints to solve for required internal forces.

Problem Statement:

Solve for the forces in members AB, AE, FE, and EC, and state if each member is in tension or compression.

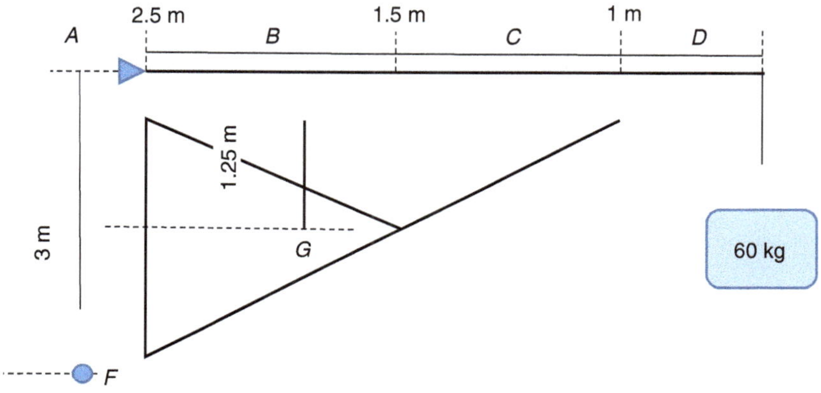

Goal:

Solve for the specified members

Given:

The system shown

Equations Needed:

$$\sum \boldsymbol{F} = 0$$

$$\sum M = 0$$

Answer:

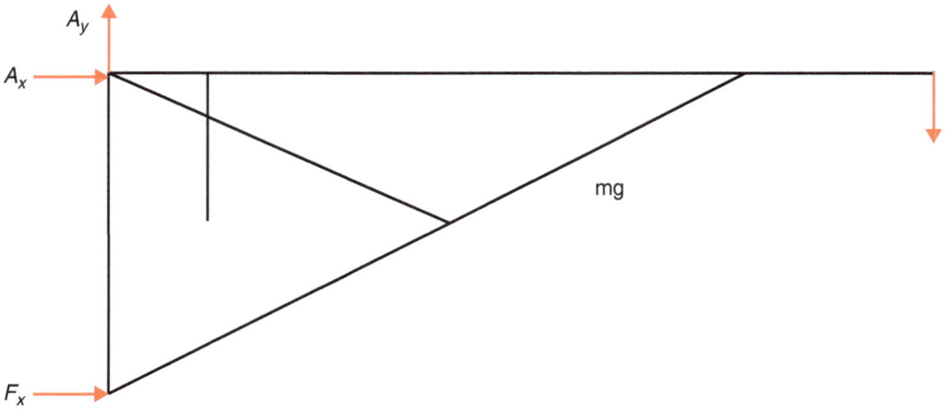

$$\sum \boldsymbol{F}_y = 0 = A_y - \text{mg} = A_y - 60(9.81)$$

$$A_y = 588.6\,\text{N}$$

$$\sum M_A = 0 = 3\boldsymbol{F}_x - 5\,\text{mg} = 3\boldsymbol{F}_x - 5(60)(9.81)$$

$$\boldsymbol{F}_x = 981\,\text{N}$$

$$\sum \boldsymbol{F}_x = 0 = A_x + \boldsymbol{F}_x = A_x + 981$$

$$A_x = -981\,\text{N}$$

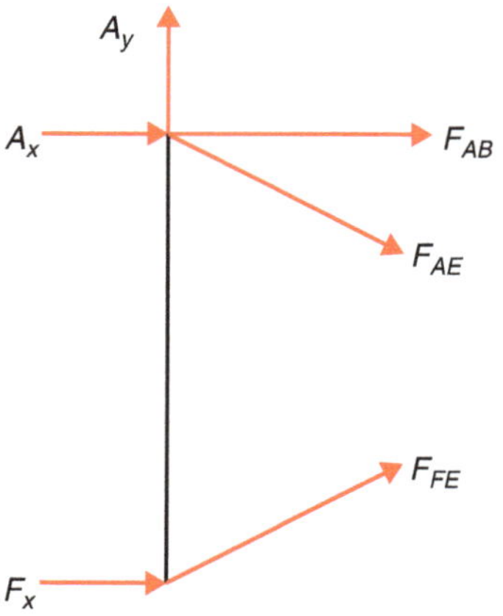

$$\sum M_A = 0 = 3F_x + 3\left(\frac{2.5}{\sqrt{2.5^2 + 1.75^2}}\right)F_{FE}$$

$$\boldsymbol{F}_{\text{FE}} = 1{,}197.46\ \text{N (C)}$$

$$\sum \boldsymbol{F}_y = 0 = A_y + \left(\frac{1.75}{\sqrt{2.5^2 + 1.75^2}}\right)\boldsymbol{F}_{FE} - \left(\frac{1.25}{\sqrt{2.5^2 + 1.25^2}}\right)\boldsymbol{F}_{AE}$$

$$\boldsymbol{F}_{\text{AE}} = 219.36\,\text{N (C)}$$

$$\sum F_x = 0 = A_x + F_x + F_{AB+}\left(\frac{2.5}{\sqrt{2.5^2 + 1.75^2}}\right)F_{FE} + \left(\frac{2.5}{\sqrt{2.5^2 + 1.25^2}}\right)F_{AE}$$

$$F_{AB} = 1{,}177.2 \text{ N (T)}$$

Point E:

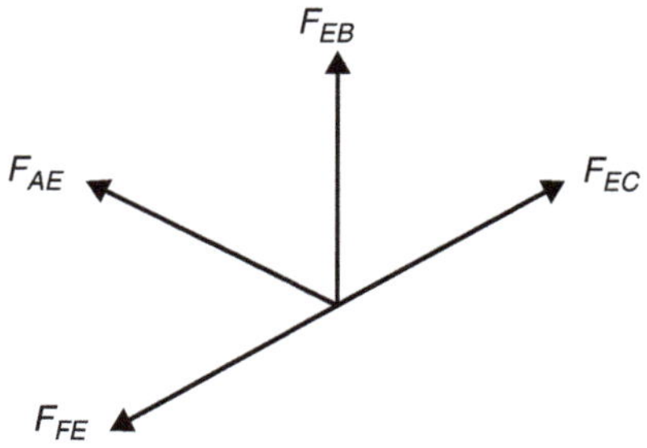

$$\sum F_x = 0 = \left(\frac{-2.5}{\sqrt{2.5^2 + 1.75^2}}\right)F_{FE} + \left(\frac{-2.5}{\sqrt{2.5^2 + 1.25^2}}\right)F_{AE} + \left(\frac{1.5}{\sqrt{1.5^2 + 1.25^2}}\right)F_{EC}$$

$$F_{EC} = 1{,}532.37 \text{ N (C)}$$

5.5. SYMMETRY OF A COMPLEX TRUSS

Problem Author: Yangling Zhou

Type of Problem:

Trusses, using the method of joints

Know how to use the method of joints to solve for the truss members and how to find the same force members by symmetry of the structure

Problem Statement:

Determine the force in each member of the loaded truss. Make use of the symmetry of the truss and of the loading.

Drawing:

If possible, please take an original photograph or draw as clear as possible the problem. Include any necessary constraints.

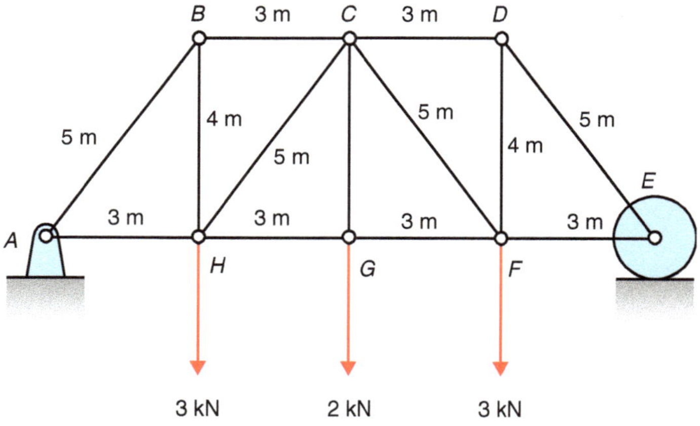

Solution:

$$F_{CD} = F_{BC} = -3\,\text{kN},\ F_{CF} = F_{HC} = -\frac{5}{4}\,\text{kN},\ F_{GF} = F_{HG} = \frac{15}{4}\,\text{kN},\ F_{DF} = F_{BH} = 4\,\text{kN},$$

$$F_{DE} = F_{AB} = -5\,\text{kN},\ F_{FE} = F_{AH} = 3\,\text{kN},\ F_{GC} = 2\,\text{kN}$$

Goal:

Find force in each member

Given:

External forces acting on the truss

Assumptions:
Truss is massless

Equations Needed:
$\sum F_x = 0, \sum F_y = 0$

Answer:
FBD:

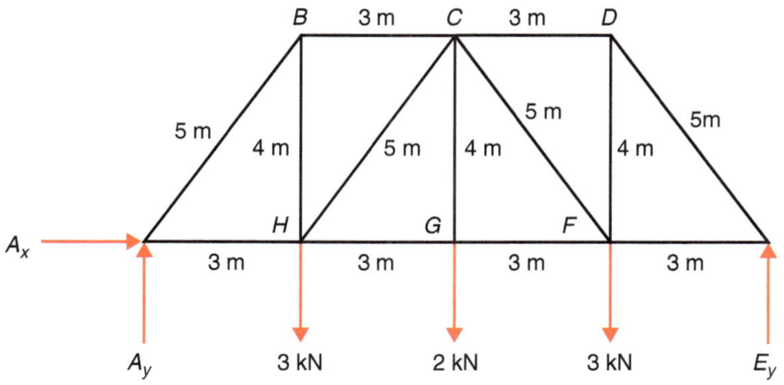

$$\sum F_x = A_x = 0$$

By symmetry, $A_y = E_y = \dfrac{3+2+3}{2}\,\text{kN} = 4\,\text{kN}$
Joint A:

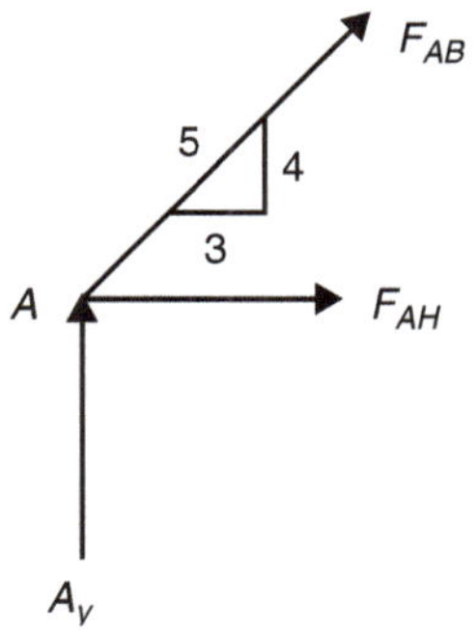

$$\sum F_y = A_y + F_{AB} \cdot \frac{4}{5} = 0 \Rightarrow F_{AB} = -5\,\text{kN} \qquad \text{Ans.}$$

$$\sum F_x = F_{AH} + F_{AB} \cdot \frac{3}{5} = 0 \Rightarrow F_{AH} = 3\,\text{kN} \qquad \text{Ans.}$$

Joint B:

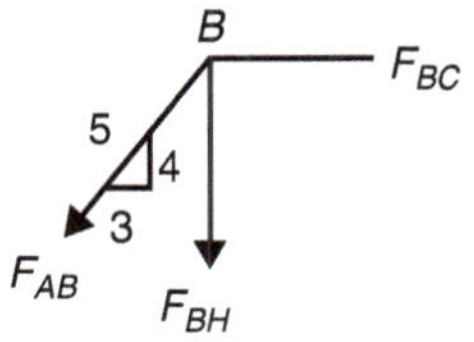

$$\sum F_x = F_{BC} - \frac{3}{5} \cdot F_{AB} = 0 \Rightarrow F_{BC} = -3\,\text{kN} \qquad \text{Ans.}$$

$$\sum F_y = -\frac{4}{5} \cdot F_{AB} - F_{BH} = 0 \Rightarrow F_{BH} = 4\,\text{kN} \qquad \text{Ans.}$$

Joint H:

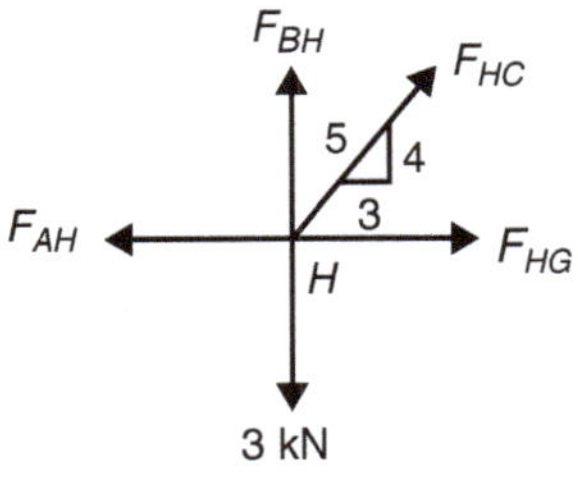

$$\sum F_y = F_{BH} + \frac{4}{5} \cdot F_{HC} - 3 = 0 \Rightarrow F_{HC} = -\frac{5}{4}\,\text{kN} \qquad \text{Ans.}$$

$$\sum F_x = -F_{AH} + \frac{3}{5} \cdot F_{HC} + F_{HG} = 0 \Rightarrow F_{HG} = \frac{15}{4}\,\text{kN} \qquad \text{Ans.}$$

Joint G:

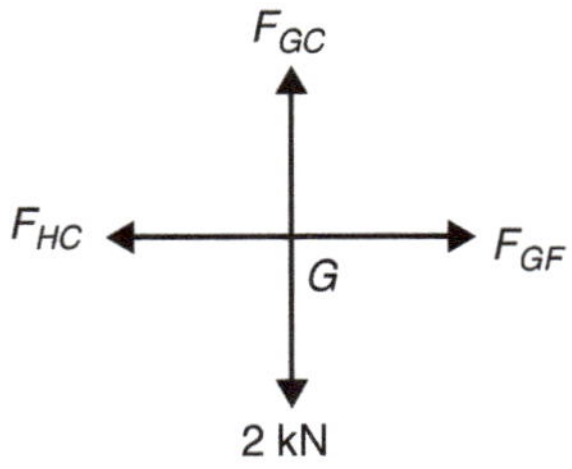

$$\sum F_y = F_{GC} - 2 = 0 \Rightarrow F_{GC} = 2\,\text{kN} \qquad \text{Ans.}$$

By symmetry:

$$F_{CD} = F_{BC} = -3\,\text{kN},\; F_{CF} = F_{HC} = -\frac{5}{4}\,\text{kN},\; F_{GF} = F_{HG} = \frac{15}{4}\,\text{kN},\; F_{DF} = F_{BH} = 4\,\text{kN},$$

$$F_{DE} = F_{AB} = -5\,\text{kN},\; F_{FE} = F_{AH} = 3\,\text{kN} \qquad \text{Ans.}$$

5.6. TRUSS ANALYSIS USING METHOD OF SECTIONS

Problem Author: Yangling Zhou

Type of Problem:
Trusses, using the method of sections

Know how to use the method of sections to solve for certain truss members

Problem Statement:

Determine the force in **CD, CF, HG** by method of sections.

Drawing:
If possible, please take an original photograph or draw as clear as possible the problem. Include any necessary constraints.

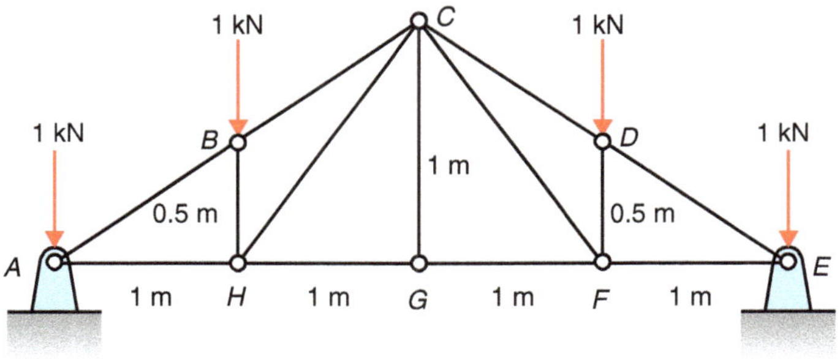

Solution:

$$F_{CD} = -2.236\,\text{kN},\ F_{CF} = 1.414\,\text{kN},\ F_{HG} = 1\,\text{kN}$$

Goal:
Find F_{CD}, F_{CF}, F_{HG}

Given:
External forces acting on the truss

Assumptions:
Truss is massless

Equations Needed:

$$\sum M = 0, \sum \boldsymbol{F}_x = 0, \sum \boldsymbol{F}_y = 0$$

Answer:

FBD of the whole truss:

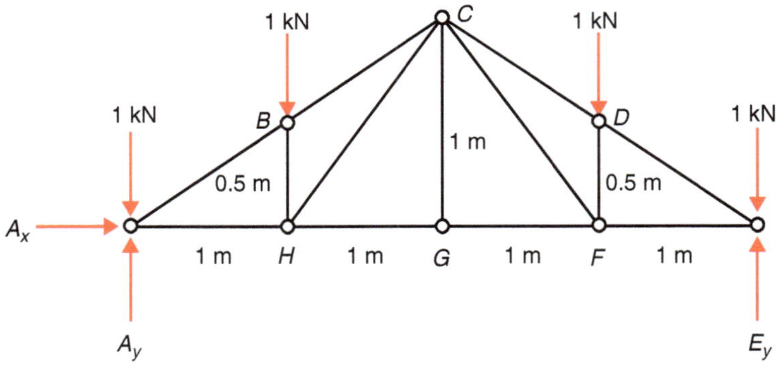

$$\sum F_x = A_x = 0$$

By symmetry,

$$A_y = E_y = \frac{1+1+1+1}{2}\,\text{kN} = 2\,\text{kN}$$

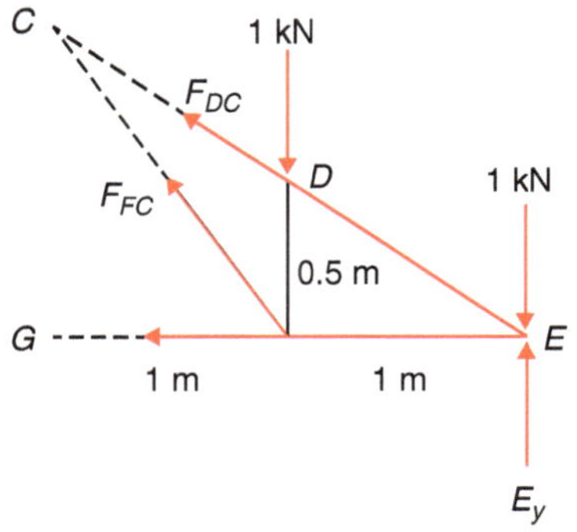

$$\tan\alpha = \frac{0.5}{1} = 0.5 \Rightarrow \alpha = 26.56°$$

$$\sum M_C = -\boldsymbol{F}_{FG}\cdot 1 - 1\cdot 1 - 1\cdot 2 + E_y\cdot 2 = 0 \Rightarrow \boldsymbol{F}_{FG} = 1\,\text{kN}$$

$$\sum M_E = -\boldsymbol{F}_{FC}\sin 45°\cdot 1 + 1\cdot 1 = 0 \Rightarrow \boldsymbol{F}_{FC} = 1.414\,\text{kN}$$ Ans.

$$\sum M_F = \boldsymbol{F}_{DC}\cos\alpha\cdot 0.5 + E_y\cdot 1 - 1\cdot 1 = 0 \Rightarrow \boldsymbol{F}_{DC} = -2.236\,\text{kN}$$ Ans.

By symmetry:

$$\boldsymbol{F}_{HG} = \boldsymbol{F}_{FG} = 1\,\text{kN}$$ Ans.

CHAPTER 6

Centroids and Distributed Loads

Distributed loads are a common application in engineering design. The forces you encounter are a type of distributed load. However, these forces have been approximated into a single point load for easy analysis. In this chapter, we will review cases when the load cannot be simplified into a point load.

Recall:
Units of distributive loads: force/length (i.e., N/m)

TOTAL FORCE OF A DISTRIBUTED LOAD

A distributed (line) load $w(x)$ is applied across a length (x) or area instead of at one point as shown in the figure. A distributed load can be equated with a concentrated load applied at a specific point along the bar.

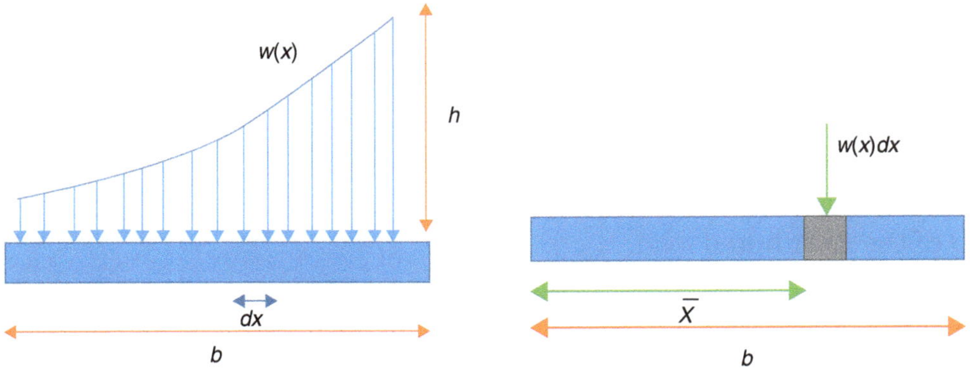

The magnitude (number) of the resultant force (concentrated) is equivalent to the area under the curve of the distributed load. We call this value the Total force of distributed load.

To calculate this:

$$\boldsymbol{F} = \int_0^b \omega(x)dx$$

ω distributed line load oriented in the y direction (force/length)
the integrand (ωdx): is the force acting on segment dx
b is the "span": length over which ω is distributed

The location of the total force (or resultant) is called the *centroid* $\bar{X}$ of the distributed line load. The moment produced by the total force should be the same that is produced by the distributed load.

To find this:

$$\bar{X} = \frac{\int_0^b x\omega(x)dx}{\boldsymbol{F}}$$

- *x is your distance from end of the load or moment center to the total force.*

Simple Shapes

If a line load distribution is a simple geometric shape. In the appendix section of your statics book you will find the derived equations for areas and centroids of these shapes.

Common examples of finding a simple distributed load:

Triangle

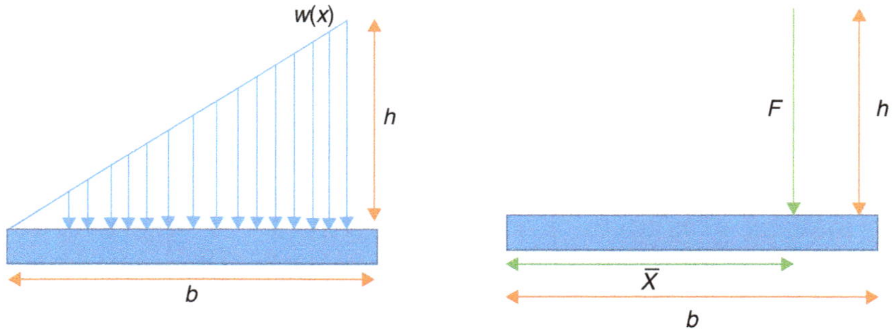

Area = Resultant (equivalent) force

$$= \frac{bh}{2}$$

Centroid of the distributed load:

$$\bar{X} = \frac{2b}{3}$$

For a triangular distributed load, the location of the resultant force is 1/3 of the length of the load, from the larger end or 2/3 from the smaller end.

Centroids

Center of Mass/Gravity

The center of mass and center of gravity are located on the same point within an object, assuming we aren't trying to do statics in a black hole (in which case, you might want to find a way out).

In studying an object, the **total mass** is the sum of the entire particle masses of the object. To location of the total mass, is the **center of mass.** This point exhibits the same behavior (or equivalent) to all the particles in the object.

For example, consider a sheet of paper.

The center of mass (the yellow arrow) is directly in the center of the paper (length and width-wise). If you were to crush the paper into a very small ball (the black dot), now all the mass is (almost) contained in a single point. This point represents the center of mass.

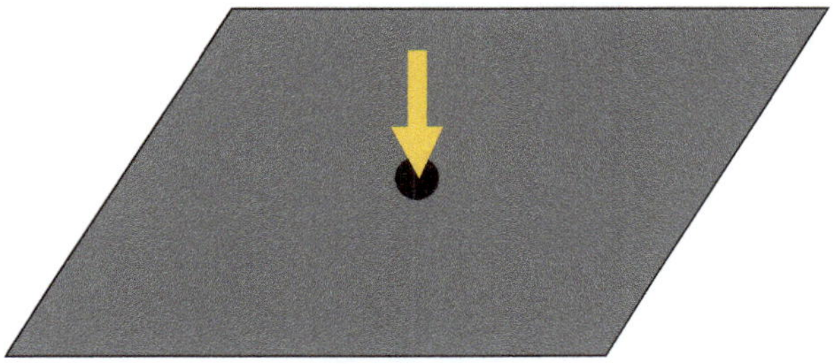

Another way to think of it, is where the material has constant density and the forces are balanced (like balancing a broom on your finger).

Location of center of mass:

Important note: Center of mass describes the physical property of an object where the centroid is the geometric property. Additionally, when your object is of constant thickness and uniform density, the centroid and center of mass occupy the same point.

$$\tilde{x} = \frac{\int x dm}{dm} \quad \tilde{x} = \frac{\int y dm}{dm} \quad \tilde{x} = \frac{\int z dm}{dm}$$

The differential element:

$$dm = pdv$$

Additional note: Center of mass we donate with a tilde where centroid we use a bar.

Centroids of Volumes and Areas

The centroid of an object is often the same as the center of mass, but only if the object's density is constant. The centroid, which is the geometric center of the object, is the physical center of the object. The centroid does not depend on the density or mass across an object. For symmetrical objects, the centroid lies on the axis of symmetry.

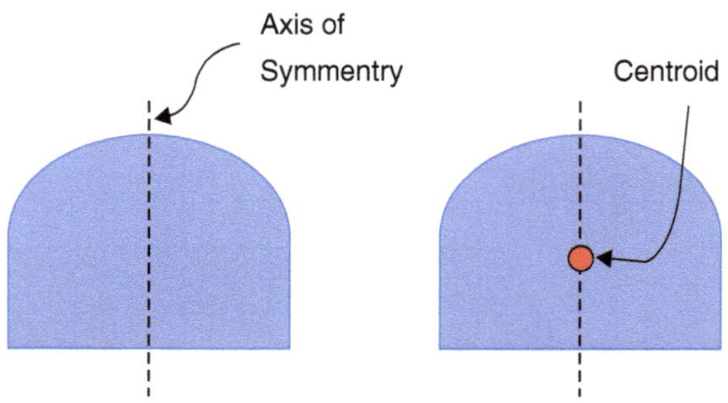

For three dimensional objects the centroid lies on planes of symmetry. If the object has two planes of symmetry, then the centroid lies on the intersection of the two planes, which is the axis of symmetry.

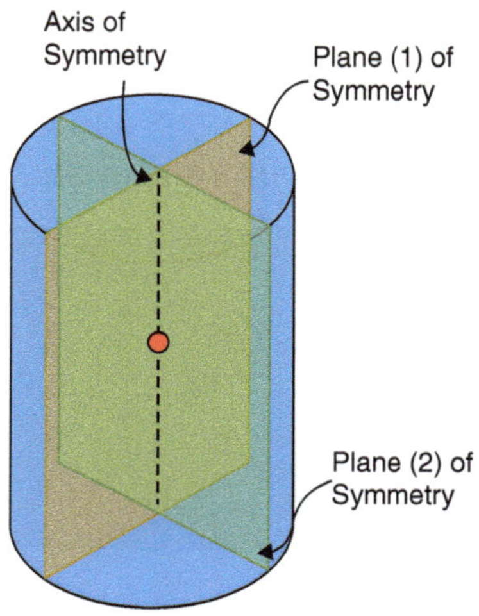

Calculation of Volume and Area

Centroid of volume:

$$\bar{x} = \frac{\int x dv}{dv} \quad \bar{y} = \frac{\int y dv}{dv} \quad \bar{z} = \frac{\int z dv}{dv}$$

Centroid of area

Assuming constant thickness (2D)

$$\bar{x} = \frac{\int x dA}{dA} \quad \bar{y} = \frac{\int y dA}{dA}$$

Center of an Area Examples

In statics, many centroid problems you'll encounter are in two dimensions. There are several shapes that are more common than others. For these common shapes, the location of the centroid is already known (usually in the appendix of a book). Two of the most common examples are rectangles and triangles.

The centroid of a triangle is located exactly one-third of the length from the tall edge of the triangle and one-third of the height from the long edge of the triangles.

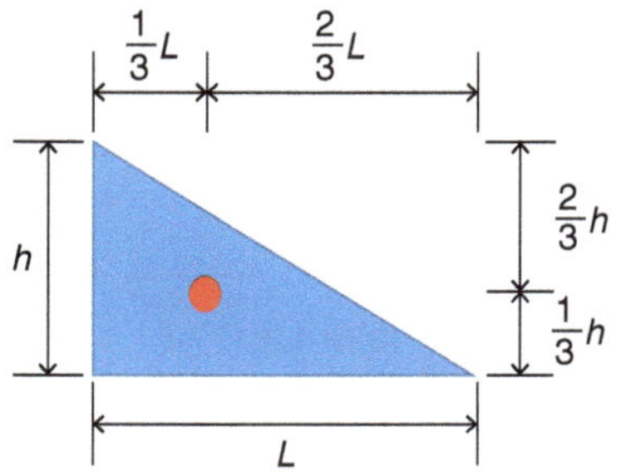

Rectangles are also common in distributed loads. The centroid of a rectangle is located exactly in the center of the length and height, meaning it is halfway between all of the sides.

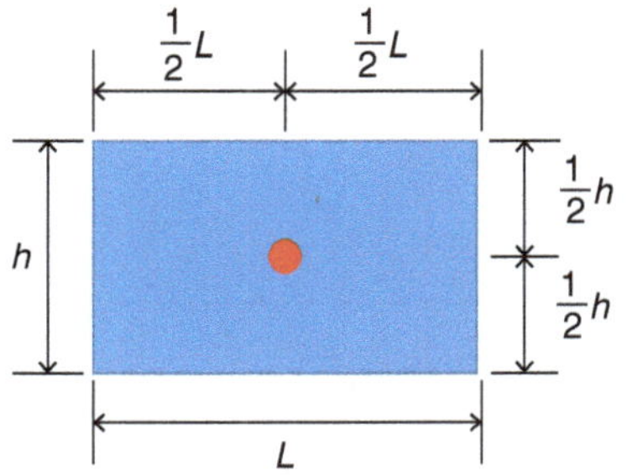

Composite shapes, how to locate the centroid $\bar{x}$ $\bar{y}$ (2D shape):

Note: *The following may also be used for volumes and center of mass.*

$$\bar{x} = \frac{\sum x_i A_i}{\sum A_i} \quad \bar{y} = \frac{\sum y_i A_i}{\sum A_i}$$

1. Divide the composite into simple shapes.
2. Determine a reference axis for the entire composite shape (x, y).
3. Calculate the area of each shape (A_i). Then calculate the entire composite area $\sum A_i$.
4. Using either integration or the predetermined equations to calculate each shapes centroid.
5. Using the reference axis, determine the distance from each shapes centroid (x_i, y_i).
6. Multiply each the area's by it corresponding distance from the x and y axis of reference $x_i A_i$ and $y_i A_i$. Then sum all the products $\sum x_i A_i$ and $\sum y_i A_i$. (If there is a hole make sure to subtract this from the shape).
7. Finish by finding $\bar{x}$, $\bar{y}$.

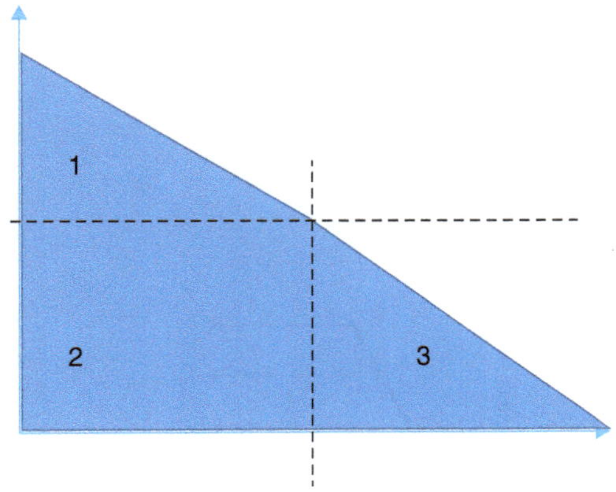

Note: The following table is shown to help organize and solve the problem.

Shape	Area	$x_i A_i$	$y_i A_i$	$\bar{x}$	$\bar{y}$
1					
2					
3					
Sum	$\sum A_i$	$\sum x_i A_i$	$\sum y_i A_i$		

Hydrostatics

Hydrostatic pressure is simply the amount of "force" that a fluid exerts on a submerged body. Hydrostatic pressure is a distributed force. The amount of pressure is directly proportional to the depth of the body in the fluid. The deeper the body is in the fluid, the more pressure it must withstand.

Note: If the surface is horizontal the loading will be uniform.

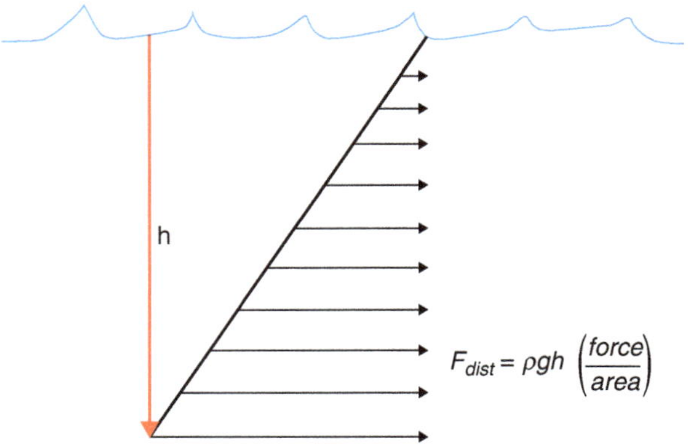

The same equation holds true for all fluid. As shown in the figure, ρ is the fluid density, h is the depth of the body within the fluid, and g is the gravitational constant. It can be seen both graphically and mathematically in the figure that the pressure exerted on the body is directly proportional to the depth of the object in the fluid.

Static fluid pressure doesn't depend on the shape, total mass, or surface area of the liquid. Therefore:

$$Pressure = \frac{Weight}{Area} = \frac{mg}{A} = \frac{\rho V g}{A} = \rho g h$$

Methodology to solve for Hydrostatics:

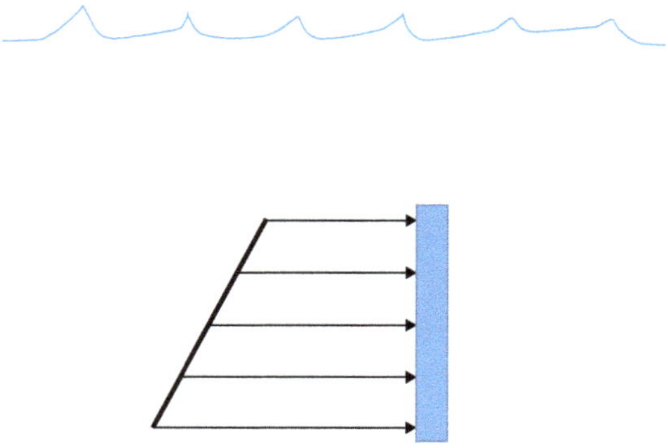

For an object submerged in the fluid, the forces will act the same as previously discussed distributed loads. As seen to the left, the distributed loading can be simplified into a rectangle and a triangle, as shown previously. Since pressure is a force across an area, you must know how wide the object is or solve for it based on a unit width. In the case of the figure shown, the force exerted on the object can be simplified into a rectangular and a triangular load.

Problems of this type are included in the chapter.

SECTION 6 Problems

6.1. CENTER OF MASS OF THE HUMAN BODY

Problem Author: Richard Blorstad

Type of Problem:
Distributed forces. This provides an example of a distributed load compared to a point load and how center of mass/gravity is important.

Problem Statement:
Consider a tight-rope and a balance beam. Which do you think is easier to walk on, the tight-rope of the balance beam?
Now, let's solve it.

1) Determine the centroid of the "person" shown in the following figure. The body of the person has been simplified to make this easier to solve.
2) Draw the free body diagram (FBD) of the person for both standing on a tight rope and a balance beam.
3) Is the person balanced on the tight-rope? What about on the balance beam? If not, how much would the person have to shift their weight in order to become balanced?

Drawing:

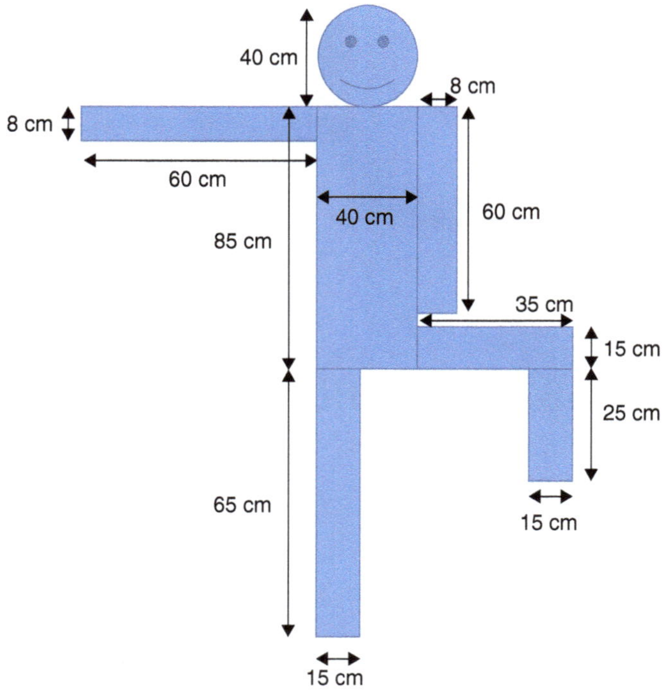

Solution:

1) $\bar{x} = 81.7\,\text{cm}$

$\bar{y} = 106.3\,\text{cm}$

2)

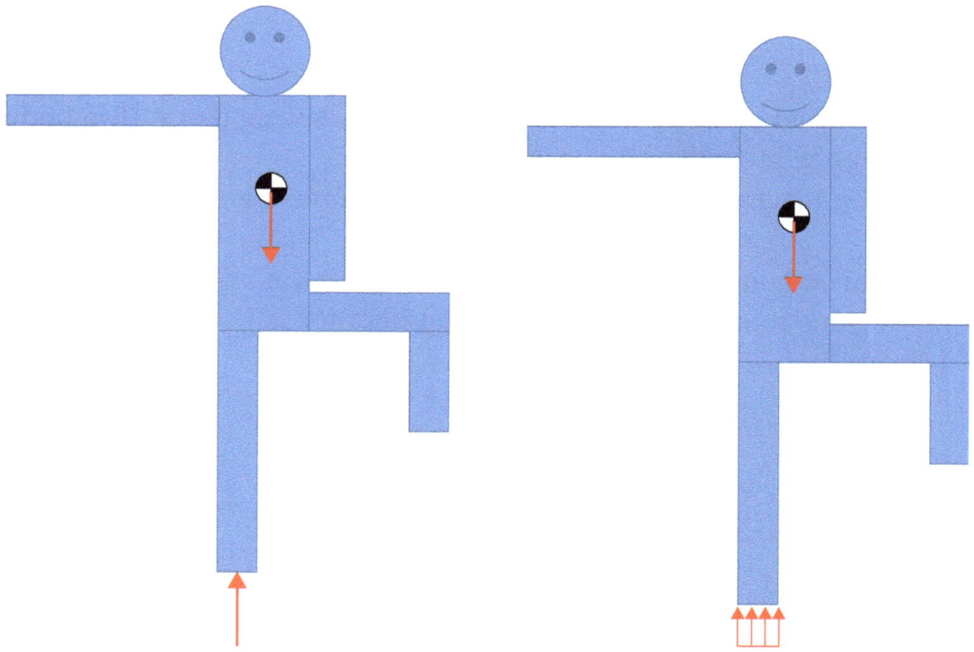

3) In both cases, the person is not balanced, but by shifting his/her weight to the left (his/her right), the person can become balanced. The required shift to balance on the tight rope is a whopping 14.2 cm, since the weight must be directly over the point load, but only needs to shift 6.7 cm to be balanced over the balance beam. Note that the distributed load is not defined and therefore could change, but as long as the weight is over the distributed load, it can be balanced.

Goal:
Solve for the center of mass of the person and determine whether or not the person can be in equilibrium in each case

Given:
Dimensions of the person

Assumptions:
Homogeneous mass throughout the person

Equations Needed:

$$\bar{y} = \frac{\sum A_i \overline{y_i}}{\sum A_i} \quad \bar{x} = \frac{\sum A_i \overline{x_i}}{\sum A_i}$$

Answer:
To solve for the centroid, the following table can be used, in accordance with the shape numbering provided.

Shape number	A_i	$\overline{y_i}$	$A_i\overline{y_i}$	$\overline{x_i}$	$A_i\overline{x_i}$
1	975	32.5	31,687.5	67.5	65,812.5
2	375	52.5	19,687.5	127.5	47,812.5
3	525	72.5	38,062.5	117.5	61,687.5
4	3,400	107.5	365,500	80	272,000
5	480	120	57,600	104	49,920
6	480	146	70,080	30	14,400
7	1,257	170	213,690	80	100,560
Sum	7,492		796,307.5		612,192.5

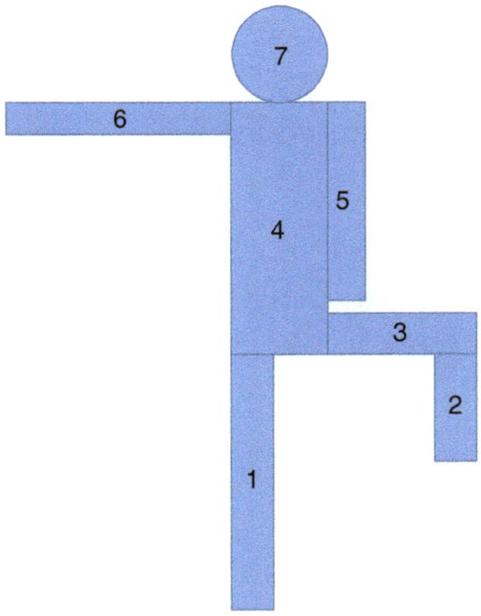

6.2. FORCE OF SNOW ON ROOF

Problem Author: Megan Centafont

Type of Problem:
This problem will cover distributed load, which involves finding the magnitude resultant force, the produced moment, and the centroid.

Problem Statement:
Snowfall exerts a distributed load on the overhang of a roof, which acts like a cantilever beam. Determine the magnitude and location of the equivalent resultant force of this load.

Drawing:

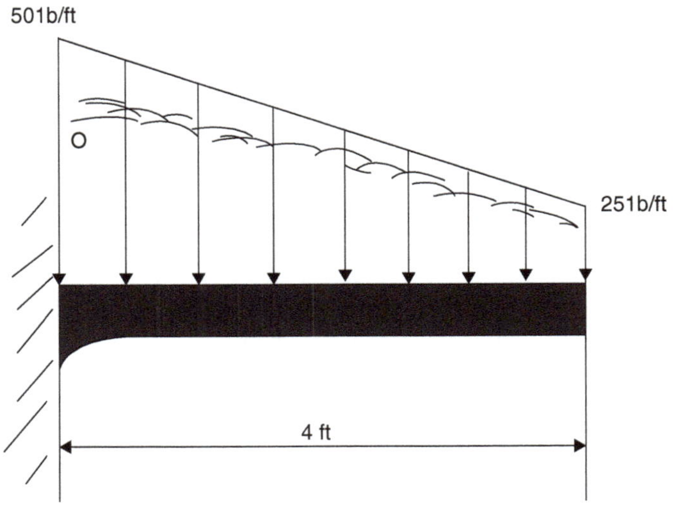

Goal:
To gain knowledge of calculating distributed loads.

Given:
Range of distributed load of 50–25 lb./ft. Length is 4 ft.

Assumptions:
Downward force and clockwise moments are *positive orientation*.

Equations Needed:
Centroid formula for a trapezoid or area for a triangle and rectangle.

Solution:

Since centroid formulas for a trapezoid are not so easily memorized, the given formation was divided into a rectangular and triangular loading and solved for by using composite areas.

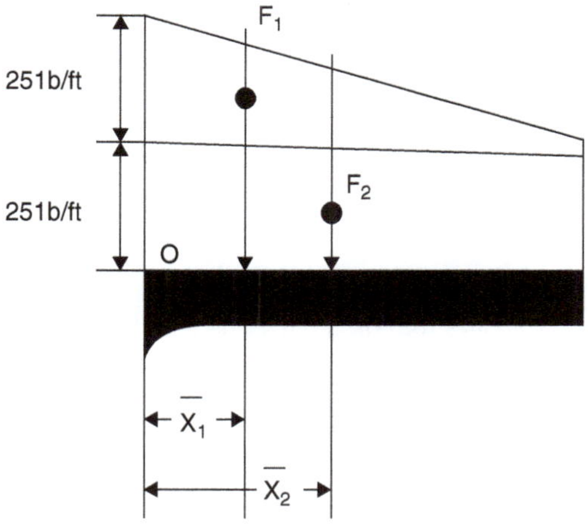

The magnitude of the force represented by each of these loadings is equal to its area.

$$F_1 = \frac{1}{2}(4\,\text{ft.})\left(\frac{25lb}{ft}\right) = 50\,\text{lb.}$$

$$F_2 = (4\,\text{ft.})\left(\frac{25\,\text{lb.}}{\text{ft.}}\right) = 100\,\text{lb.}$$

These parallel forces act through the respective centroids of their areas and therefore intersect the beam at,

$$\bar{x}_1 = \frac{1}{3}(4\,\text{ft.}) = 1.33\,\text{ft.}$$

$$\bar{x}_2 = \frac{1}{2}(4\,\text{ft.}) = 2\,\text{ft.}$$

Combing the forces into a single resultant force,

$$+\downarrow F_R = \sum F$$

$$F_R = 50 + 100 = 150\,\text{lb.}$$

To find the location of the total resultant force,

$$+\circlearrowleft (M_R)_O = \sum M_R$$

$$\bar{x}(150) = 1.33(50) + 2(100)$$

$$\bar{x} = 1.78\,\text{ft.}$$

6.3. EQUILIBRIUM (AND RISKY DECISIONS) INVOLVING BEAM ANALYSIS

Problem Author: Jordan Bendler

Type of Problem/Concept:

The concept that this problem covers is distributive loads. The problem tests students' ability to perform different calculations and conversions based on a given distributed load system. Then, using those calculations, the students are asked to determine a specific value given a hypothetical situation involving the system.

Problem Statement:

A local engineering student from Engineering Student University (ESU) takes a trip to New York City. During his trip, he visits the top of the Empire State Building. There, he finds a beam jutting out over the edge of the observation deck. The student wants to get a picture from the very edge of the beam, but first he needs to do some static analysis to make sure the beam is safe.

Drawings:

Here are diagrams of the beam:

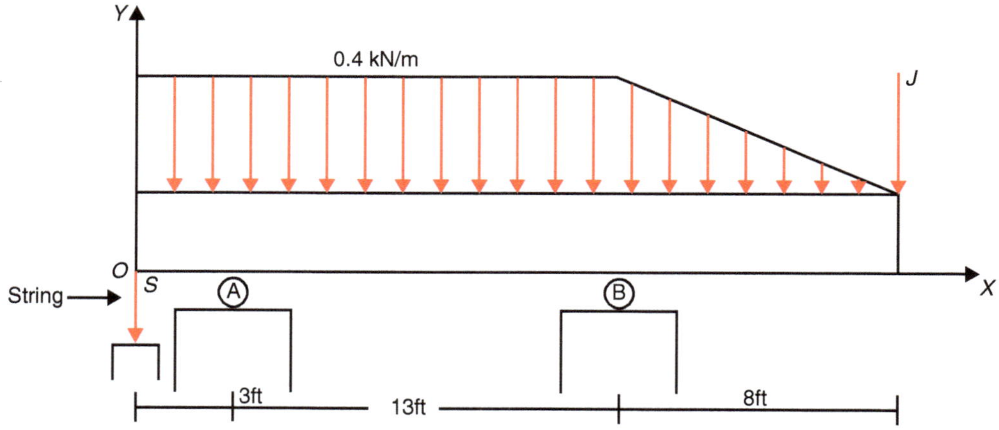

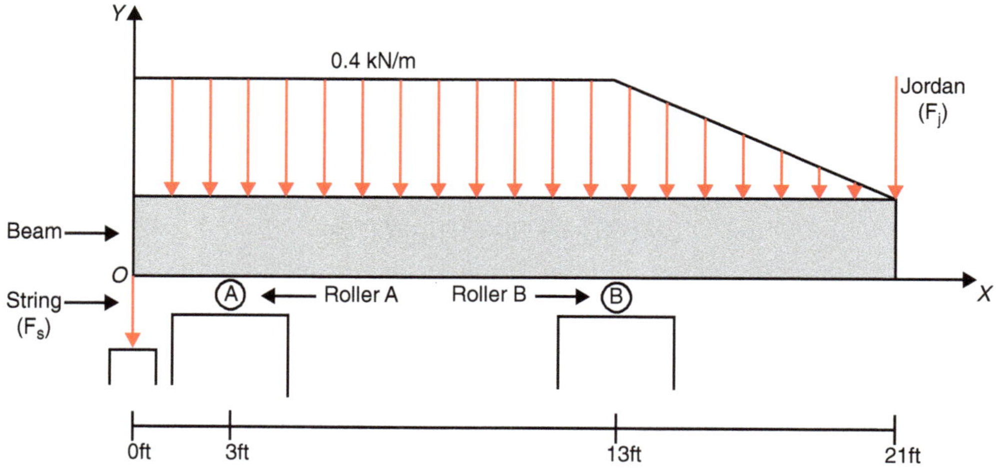

Goal:

Help the student find:

A) The equivalent load of the distributed forces

B) The forces acting at:

  a) Roller A

  b) Roller B

C) If the maximum tension that the string ($\boldsymbol{F}_s$) can hold is 0.5 kN, what is the maximum force (in newton) that the student ($\mathbf{F}_j$) can apply to the very edge of the beam before the string breaks and the student falls?

  a) If the student exerts a force of 645 N, and his camera gear adds a force of 67 N, will the string break if the student goes to the far edge of the beam?

Given:

- Diagram of beam
- Student and camera's total force (In newton)
- Distributed load forces (In newton)
- Distances (in feet)

Assumptions:

- The student steps directly into place when boarding the beam (No forces in X-direction)
- Beam is at equilibrium before the student steps on
- There is no force in the string until the student boards the beam
- In part C, when the student steps onto the edge of the beam, the beam slightly lifts off roller A ($A_y = 0$)

Equations Needed:

$$\Sigma f_x = 0 \, \Sigma f_y = 0 \;\; \Sigma M_z = 0$$

$$X_{c\text{-triangle}} = \frac{2b}{a}$$

$$\sum_{i=1}^{N} \frac{\boldsymbol{F}_i X_{ic}}{\boldsymbol{F}_{\text{total}}}$$

Solution:

The solution is on the PDF sent with this document. The name of the file is "Statics Workbook Solution.pdf."

Answer:

Part A

$$f_{\text{equivelent-rectangle}} = 1.56\,\text{kN} \quad X_{c\text{-rectangle}} = 1.95\,\text{m}$$

$$f_{\text{equivelent-triangle}} = 0.5\,\text{kN} \quad X_{c\text{-triangle}} = 4.8\,\text{m}$$

Part B

$$A_y = 0.85\,\text{kN} \quad B_y = 1.21\,\text{kN}$$

Part C

$$J_{\text{max}} = 1.798\,\text{kN}, \text{ The student will not break the string.}$$

6.4. FORCES ACTING ON A PLATE UNDERWATER

Problem Author: Kurtis Boulter

Question:
This question will help students understand the concept of hydrostatics, and will also give them practice with summing moments about a point.

Problem Statement:
A tank with a square door and a hinge at point *A* is filled with water (ρ = 1,000 kN/m). Calculate the resultant force ($\boldsymbol{F}_R$) on the door and the distance from point *B* to where the resultant force acts.

Drawing:

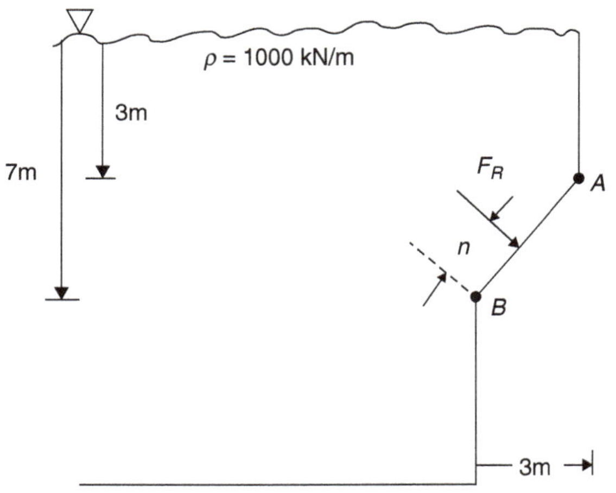

Find:
The force $\boldsymbol{F}_R$ The distance from point *B* to where $\boldsymbol{F}_R$ acts.

Given:
Location of points *A* and *B*, Various lengths (shown in drawing), density of water, 3-4-5 angle label

Assumptions:
Density is uniform, problem is taken as 2 dimensional (width is 1m), 9.81 m/s^2

Equations Needed:
Pressure is $\rho \times g \times z$, sum of the moments about *B*, area of a triangle and area of a rectangle

Solution:

1) Find pressure at A and pressure at B:

$$PA = \rho \times g \times z = 1{,}000(\text{kg/m}^3)\times 9.81(\text{m/s}^2)\times 3\,\text{m} = 29.430\text{kPa}$$

$$PB = \rho \times g \times z = 1{,}000(\text{kg/m}^3)\times 9.81(\text{m/s}^2)\times 7\,\text{m} = 68.670\text{kPa}$$

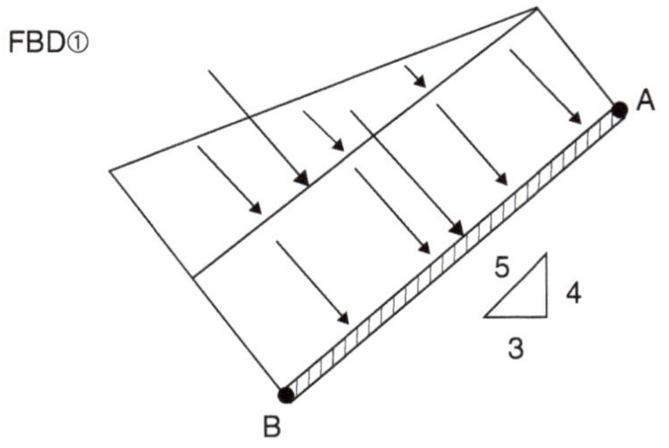

2) Find forces:

$$\boldsymbol{F}_1(\text{rectangle}) = l \times PA = 5\,\text{m} \times 29.430\ \text{kPa} = 147.150\ \text{kN}$$

$$\boldsymbol{F}_2(\text{triangle}) = \frac{1}{2}\times l \times (PB - PA) = \frac{1}{2}\times 5\,\text{m}\times 39.240\,\text{kPa} = 98.100\text{kN}$$

$\boldsymbol{F}_1$ acts at the center of the rectangle → 2.5 m from B

$\boldsymbol{F}_2$ acts at the lower third of the triangle → 1/3 × 5 = 1.667 m from B

$$F_R = F1 + F2 = 147.150\,\text{kN} + 898.100\,\text{kN} = 245.25\,\text{kN}$$

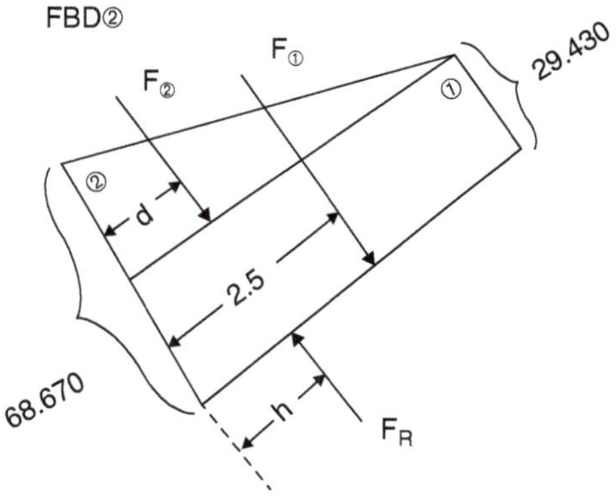

3) Sum moments about B:

$$\sum M_B = 0 = -147.15\,\text{kN}\times(2.5\,\text{m}) - 98.1\,\text{kN}\times(1.667\,\text{m}) + 245.25\,\text{kN}\times(h\,\text{m})$$

Ans: $\boldsymbol{F}_R$ = 245. 25 kN, h = 2.1646 m

6.5. KURTIS' UNDERWATER DILEMMA

Problem Author: Kurtis Boulter

Question:
This question will help students understand the concept of hydrostatics, and will also give them practice with summing moments about a point.

Problem Statement:
Kurtis is in an undersea chamber with a 1 m wide door conducting a scientific experiment. One day, the latch on the door breaks, how much force must Kurtis apply to the door's handle in order to keep it closed and prevent water from rushing in?

Drawing:

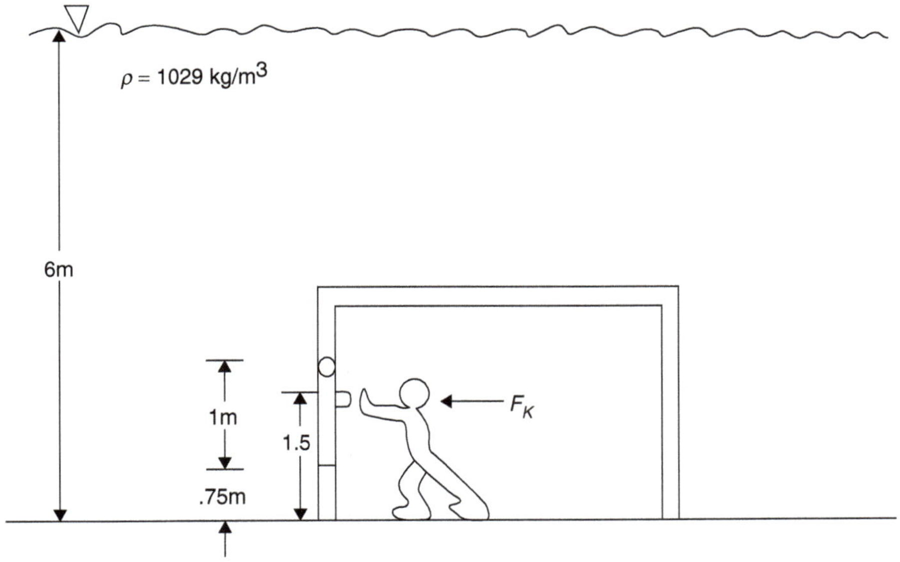

Find:
The force F_K that must be applied to keep the door shut.

Given:
Location of points *A* and *B*, Various lengths (Shown in drawing), density of seawater

Assumptions:
Density is uniform, problem is taken as 2 dimensional (width is 1m), 9.81 m/s^2

Equations Needed:

Pressure is $\rho \times g \times z$, sum of the moments about A, area of a triangle and area of a rectangle.

Solution:

1) Find pressure at A and pressure at B:

$$PA = \rho \times g \times z = 1{,}029(\text{kg}/\text{m}^3) \times 9.81(\text{m}/\text{s}^2) \times 4.25\,\text{m} = 43.319\,\text{kPa}$$

$$PB = \rho \times g \times z = 1{,}029(\text{kg}/\text{m}^3) \times 9.81(\text{m}/\text{s}^2) \times 5.25\,\text{m} = 52.996\,\text{kPa}$$

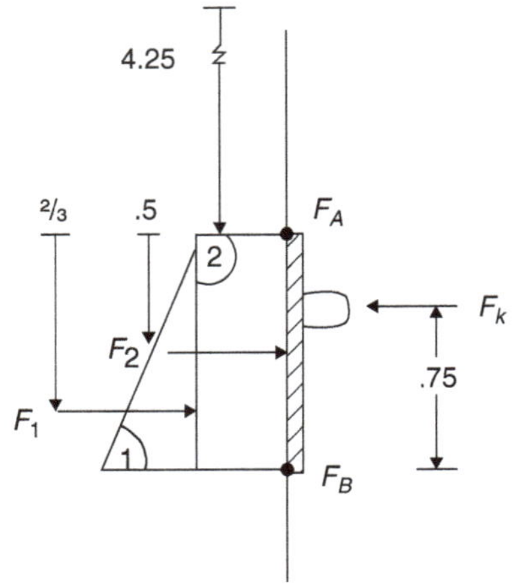

2) Find forces:

$$F_1(\text{rectangle}) = l \times PA = 1\,\text{m} \times 43.319\,\text{kPa} = 43.319\,\text{kN}$$

$$F_2(\text{triangle}) = \frac{1}{2} \times l \times (PB - PA) = \frac{1}{2} \times 5\,\text{m} \times 9.677\,\text{kPa} = 4.839\,\text{kN}$$

F_1 acts at the center of the rectangle → 1/2 m from A
F_2 acts at the lower third of the triangle → 2/3 × 5 = 1.667 m from A

$$F_R = F_1 + F_2 = 43.319\,\text{kN} + 4.839\,\text{kN} = 48.158\,\text{kN}$$

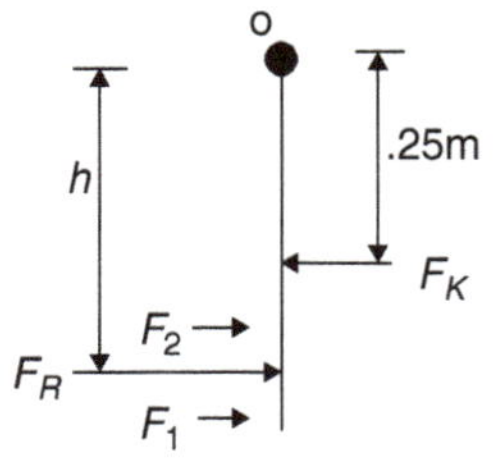

3) Sum moments about A:

$$\sum M_A = 0 = 43.319\,\text{kN} \times \left(\frac{1}{2}\text{m}\right) + 4.839\,\text{kN} \times \left(\frac{2}{3}\text{m}\right) - 48.158\,\text{kN} \times (h\,\text{m})$$

$h = 0.5167$ m

4) Sum moments about A to get F_K

$$\sum M_A = 0 = -F_K\,\text{kN} \times \left(\frac{1}{4}\text{m}\right) + 48.158\,\text{kN} \times \left(\frac{1}{2}\text{m}\right)$$

$F_K = 99.532$ kN

6.6. KURTIS' UNDERWATER SOLUTION

Problem Author: Kurtis Boulter

Question:
This question will use information from the previous question to give the students review of material they learned in a previous chapter.

Problem Statement:
Kurtis is starting to get tired of holding the door of his chamber shut and he knows he needs to get back to his experiments. Kurtis only has a length steel cable so he decides to tie it over the door handle and attach it to the wall above and below the door. If the cable can only withstand 300 kN of tension will it be adequate to hold the door shut?

Drawing:

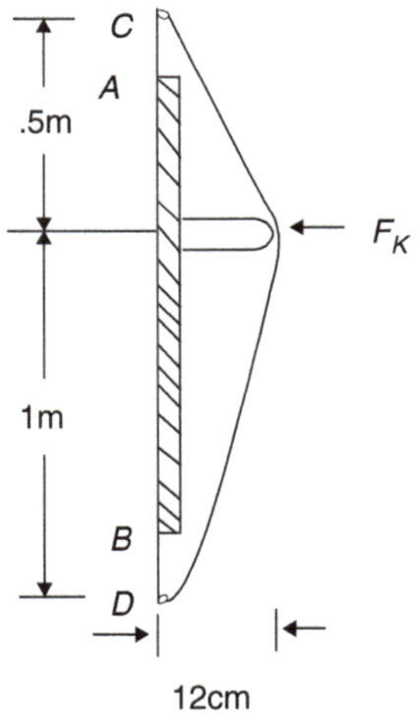

Find:
The tension in each string and compare to the maximum tension given.

Given:
Location of points C and D, Various lengths (Shown in drawing), maximum tension in the rope, door handle has a length of 12 cm

Assumptions:
Density is uniform, problem is taken as 2 dimensional (width is 1 m), 9.81 m/s^2

Equations Needed:

Law of sines, Pythagorean theorem

Solution:

1) Determine the angle between the rope and the horizontal axis

12 cm = 0.12 m

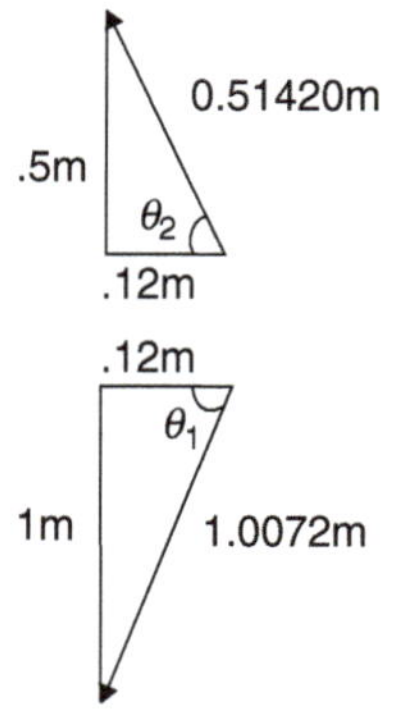

Rope (1) $\tan^{-1}\left(\dfrac{0.5}{0.12}\right) = 76.504°$

Rope (2) $\tan^{-1}\left(\dfrac{1}{0.12}\right) = 83.157°$

2) Calculate forces

We know from the previous problem that $\boldsymbol{F}_K$ is 99.532 kN, and we know that force is split between two cables so the x component of the force in each cable is:

$$\frac{\boldsymbol{F}_K}{2} = \frac{99.532\,\text{kN}}{2} = 49.766\,\text{kN}$$

$$\frac{49.766}{\sin(90-76.504)} = \frac{\boldsymbol{F}_1}{\sin(90)} \rightarrow \boldsymbol{F}_1 = 213.243\,\text{kN}$$

$$\frac{49.766}{\sin(90-83.157)} = \frac{\boldsymbol{F}_2}{\sin(90)} \rightarrow \boldsymbol{F}_2 = 417.648\,\text{kN}$$

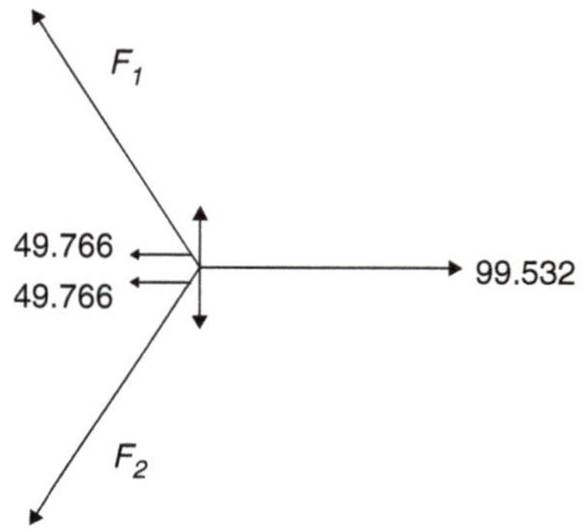

Ans: The cable will not withstand the tension; the tension in the upper cable is below the limit but the tension in the lower cable is greater than the limit so the cable will fail.

6.7. FORCES ACTING ON A SUBMERGED BODY

Problem Author: Kurtis Boulter

Question:
This question will help students understand the concept of hydrostatics, and will also give them practice with summing moments about a point.

Problem Statement:
Find the resultant force on three of the sides of the submerged vessel if ρ = 62.4 slug/ft^3.

Drawing:

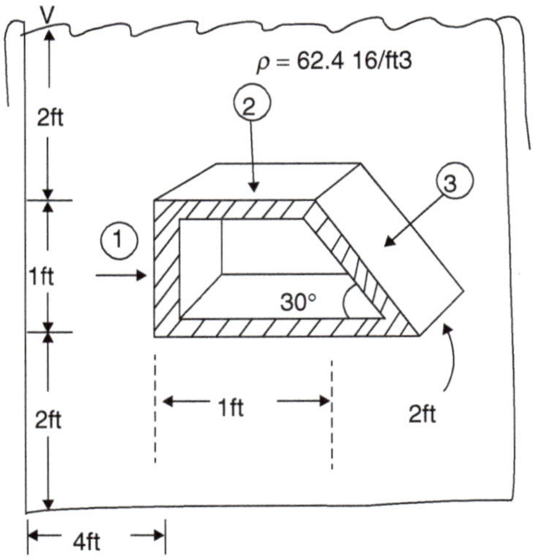

Find:
The resultant force on each side.

Given:
Various lengths and depths (Shown in drawing), density of water

Assumptions:
Density is uniform, problem is taken as 3 dimensional (width is 2 m), 32.2 ft./s^2

Equations Needed:
Pressure is $\rho \times g \times z$, sum of the moments about A, area of a triangle and area of a rectangle,

Solution:

1) Side 1 (left side)

$$PA = \rho \times g \times z = 62.4(\text{slug}/\text{ft}^3) \times 32.2(\text{ft}/\text{s}^2) \times 2\,\text{m} = 4{,}028.56\,\text{lbf}$$

$$PB = \rho \times g \times z = 62.4(\text{slug}/\text{ft}^3) \times 32.2(\text{m}/\text{s}^2) \times 3\,\text{m} = 6{,}027.84\,\text{lbf}$$

$$\boldsymbol{F}_1(\text{rectangle}) = l \times w \times PA = 1 \times 2 \times 4{,}018.56 = 8{,}037.12\,\text{slugs} @ 0.5\,\text{ft.}$$

$$\boldsymbol{F}_2(\text{triangle}) = \frac{1}{2} \times l \times w \times (PB - PA) = \frac{1}{2} \times 1 \times 2 \times 2{,}009.28 = 2{,}009.28\,\text{slugs} @ \frac{2}{3}\,\text{ft.}$$

$$\sum M_A = 0 = 8{,}037.12 \times \left(\frac{1}{2}\right) + 2{,}009.28 \times \left(\frac{2}{3}\right) - 10{,}046.40 \times (h) \rightarrow d = h + 2$$

Ans: (side 1): d = 2.533 ft., $\boldsymbol{F}_R$ = 10,046.40 slugs

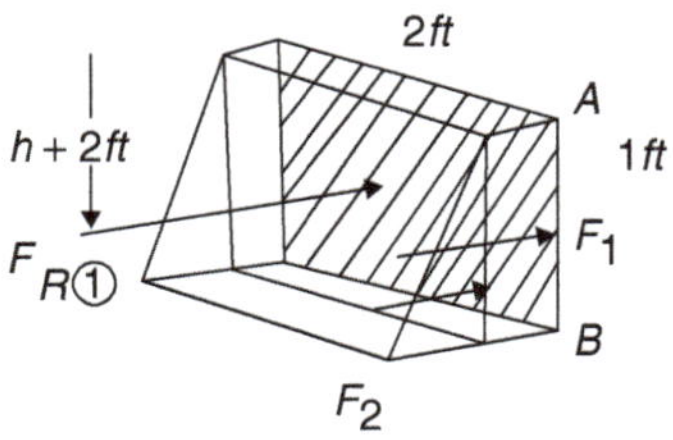

2) Side 2 (top side):

$$P = \rho \times g \times z = 62.4(\text{slug}/\text{ft}^3) \times 32.2(\text{ft.}/\text{s}^2) \times 2\,\text{m} = 4{,}028.56\,\text{lbf}$$

$$\boldsymbol{F}_1(\text{rectangle}) = l \times w \times PA = 1 \times 2 \times 4{,}018.56 = 8{,}037.12\,\text{slugs}$$

Ans: (side 2): d = 2 ft., $\boldsymbol{F}_R$ = 8,037.12 slugs

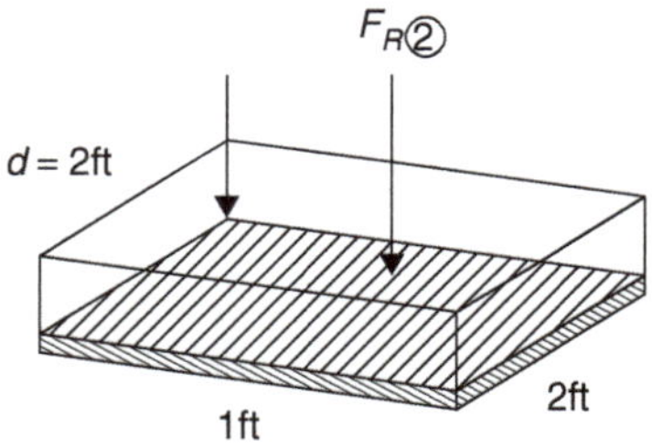

3) Side 3 (right side):

Find hypotenuse:

$$\tan(30) = \frac{1}{l} \rightarrow l = 1.732$$

$$\text{hypotenuse} = \sqrt{1^2 + 1.732^2}$$

From part 1 $\rightarrow \boldsymbol{F}_A$ = 4,018.56 lbf, $\boldsymbol{F}_B$ = 6,027.84 lbf

$$\boldsymbol{F}_1(\text{rectangle}) = l \times w \times PA = 2 \times 2 \times 4{,}018.56 = 16{,}074.24\,\text{slugs} @ 1\,\text{ft. from } A$$

$$F_2(\text{triangle}) = \frac{1}{2} \times l \times w \times (PB - PA) = \frac{1}{2} \times 1 \times 2 \times 2{,}009.28 = 2{,}009.28\,\text{slugs} @ \frac{4}{3}\text{ ft. from } A$$

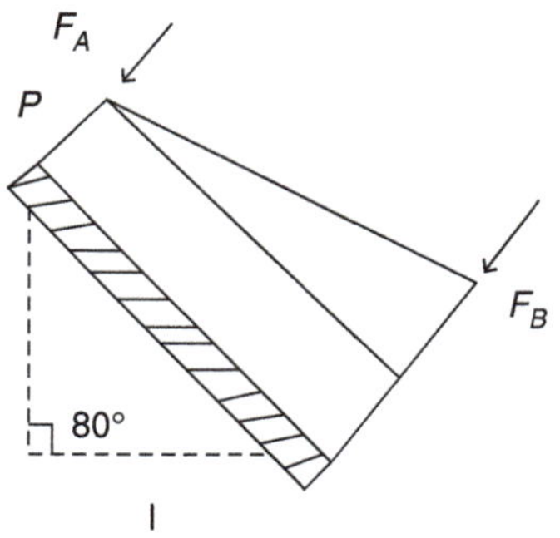

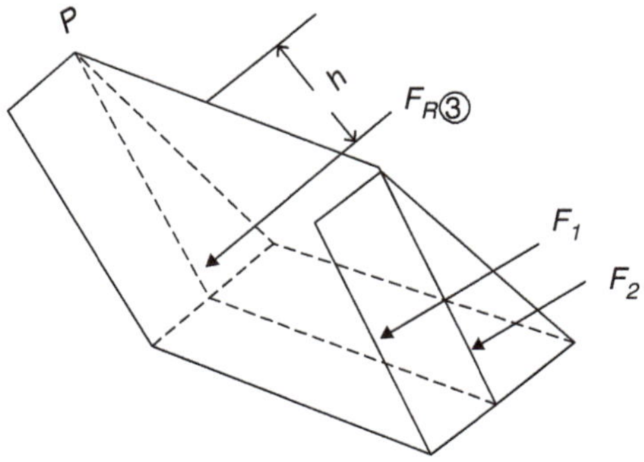

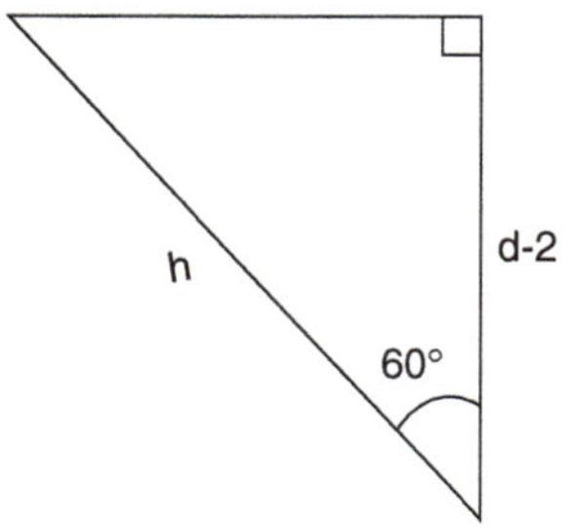

6.8. CENTROID OF A COMPOSITE SHAPE

Problem Author: Prof. Jamie Gurganus

Type of Problem:
Centroid of the composite shape.

Question:
Find the centroid of the composite shape shown.

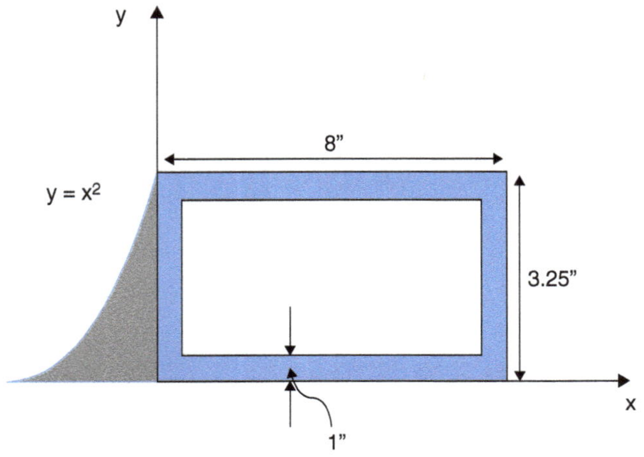

CHAPTER 7

Shear and Bending Moment Diagrams

In design analysis, it's necessary to locate where the maximum and minimum loads are in a structure, especially in beam analysis. In this chapter, a review of method sections and shear and bending moment diagrams are discussed.

Definition and types

A beam is a structural element subjected to moments and forces acting on the perpendicular axis.

Note: *A two-force member is not a beam because there are no loads acting perpendicular to its long axis.*

Five types of statically determinate beams:

Simply supported beams
Simply supported with an overhang beams
Cantilever beams
Balancing beams
Fixed—Fixed

Beam internal loads (axial, shear, moment)

Recall: When a load is applied to a beam the moment, which is internal to a beam, is called a bending moment. The force in the plane of the cross section of the beam is called a shear force, and the force perpendicular to the cross section is called an axial (normal) force. (See diagram on the following page)

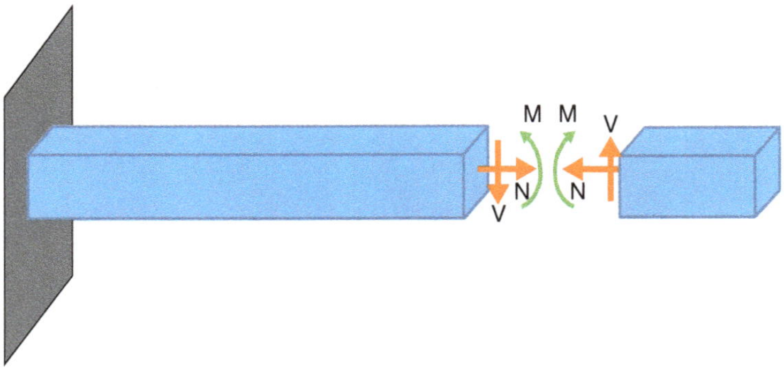

PROCEDURES FOR FINDING INTERNAL LOADS IN BEAMS

Method of Sections

To determine the amount of sections to create in a beam will depend on the beam loading and configuration.

The beam must be sectioned at each point that loading or support conditions change. A cut must be made in each section of the beam so that the shear and moment can be evaluated over that section.

Step I: Determine support reactions

Step II: Draw the free-body diagram (FBD) of the selected portion of the beam

Step III: Determine the values for the shear force and bending moment

Multiple sections are required since shear and bending moments are not a continuous function over changing supports and loads.

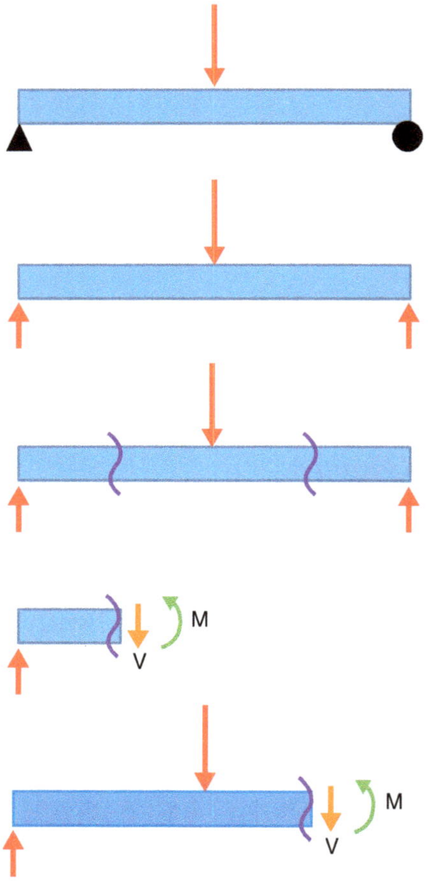

SHEAR AND BENDING MOMENT DIAGRAMS

A tool to determine maximum or minimum forces along a beam are shear and bending moment diagrams. There is a relationship that exist between the loading, shear, and bending moments. Determining the area of the loading gives the change in shear (V) force and the area of the shear gives the change in bending moment (M).

$$w(x) = \text{the load function}$$

$$V(x) = \int w(x)\,dx$$

$$M(x) = \int V(x)\,dx$$

Procedure for Creating Shear and Bending Moment Diagrams

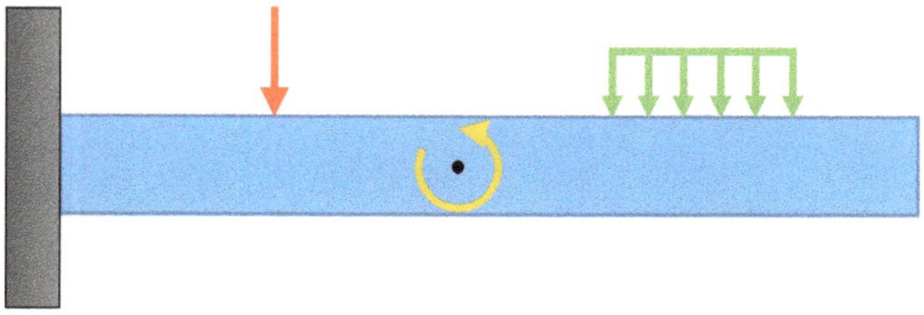

Step I: Determine the forces on the support reactions. To solve for support reactions, you will need to draw an FBD of the entire beam and utilize the Equations of Equilibrium.

Note: In the example, axial force is not shown because it equals zero.

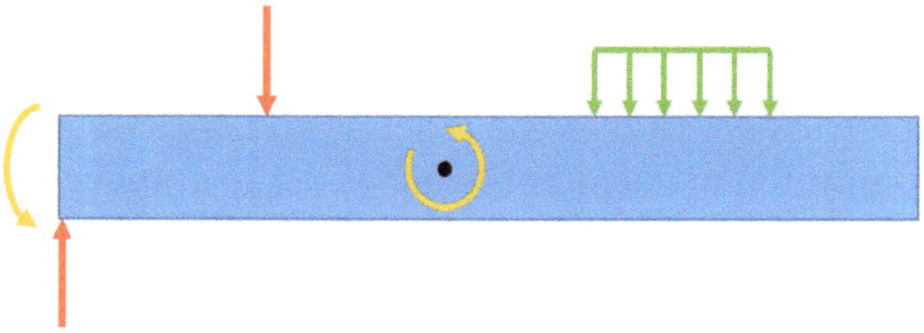

Step II: Identify points where the shear force and bending moment must be computed. Section the beam according to the identified points.

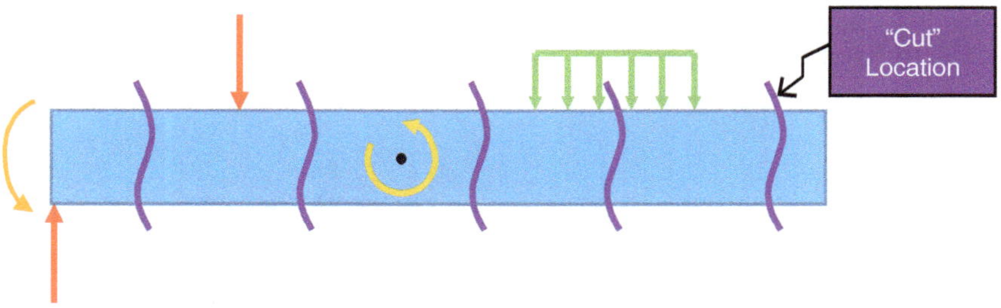

Step III: At each segment, use the equations of equilibrium to find the shear force and bending moment.

Note: Make sure to draw the FBD of each of the segments. Be sure shear (V) and bending moment (M) are shown acting in their *positive sense (sign convention).*

Recall that the magnitudes of the shear force and bending moment may be functions of x.

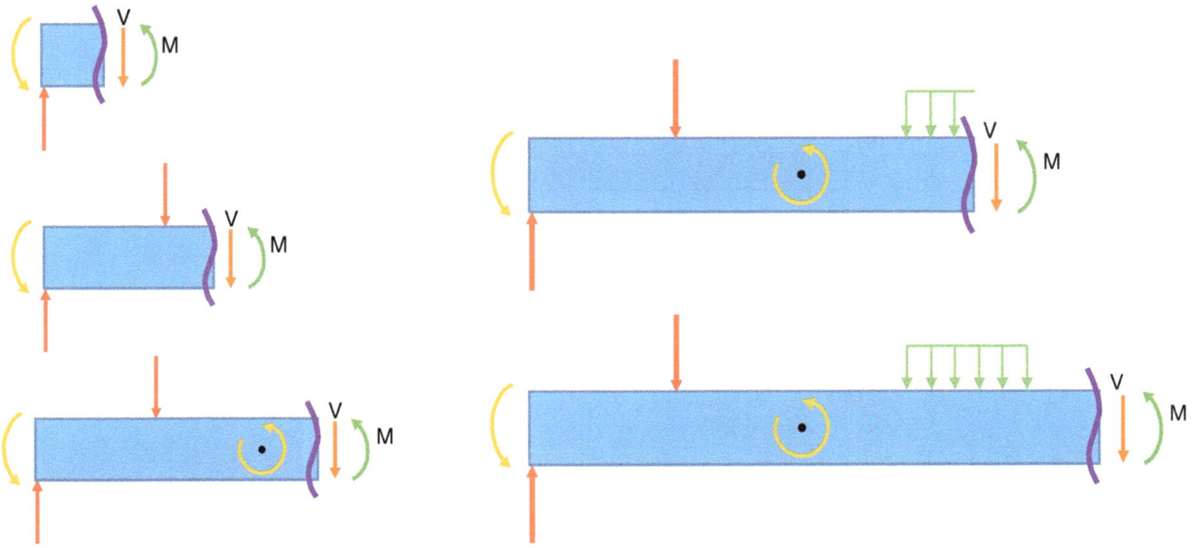

Step IV: Draw the shear force diagram.

Create a coordinate axes with x as the abscissa and V as the ordinate. Proceed from the left end. Start with zero shear force. Using the shear force equations for each segment, plot the shear force diagram.

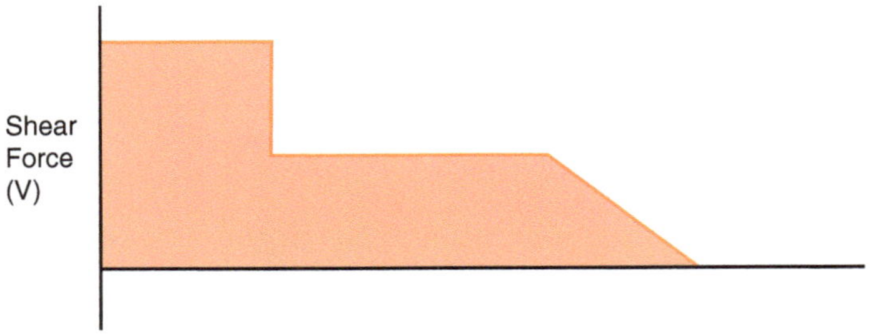

Step V: Draw the bending moment diagram.

Establish a set of coordinate axes with x as the abscissa and M as the ordinate. Proceed from the left end. If there is a concentrated moment at the left end, the graph will jump at zero. If there are no concentrated moments acting on the beam, the bending moment diagram will be continuous and it will start and end at zero.

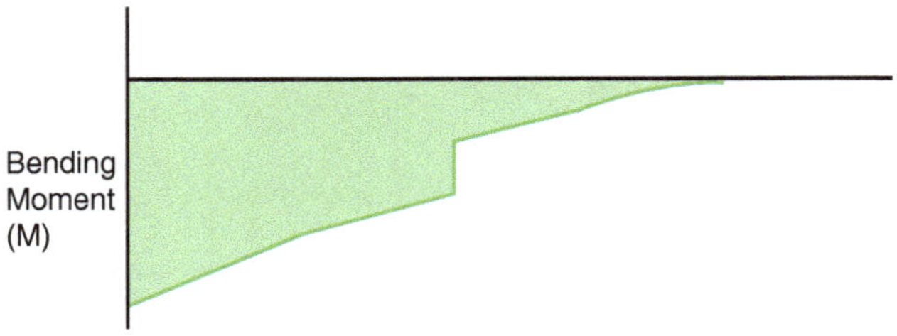

Note: It is often simpler to include the FBD, shear force diagram, and the bending moment diagram all together in line with one another. This can help you better understand where the changes in the diagrams should be.

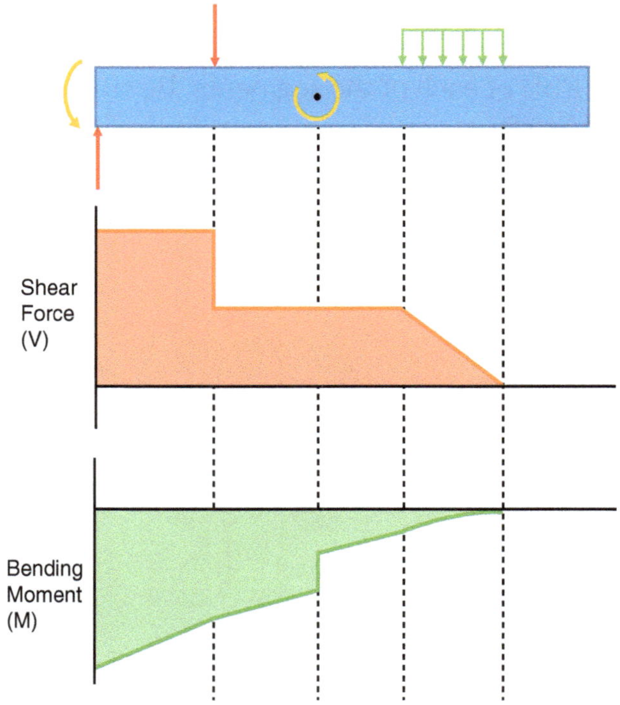

SECTION 7

Problems

7.1. *V* AND *M* DIAGRAMS OF A BEAM

Problem Author: Sarah Hemler

Type of Problem:
This problem will address shear in a beam. The student should be able to solve for the external loads on the beam and calculate the internal shear force in the beam at any point.

Problem Statement:
In order to reach a party on a lakefront deck, attendees must walk along a 10 m wooden deck supported at either end. At any point along the walkway, the shear force may not exceed 6 kN. If people are lined up along the walkway simulating a distributed load of 1.2 kN/m as shown in the following figure, what is the maximum shear the beam will experience? Will the beam break?

Drawing:

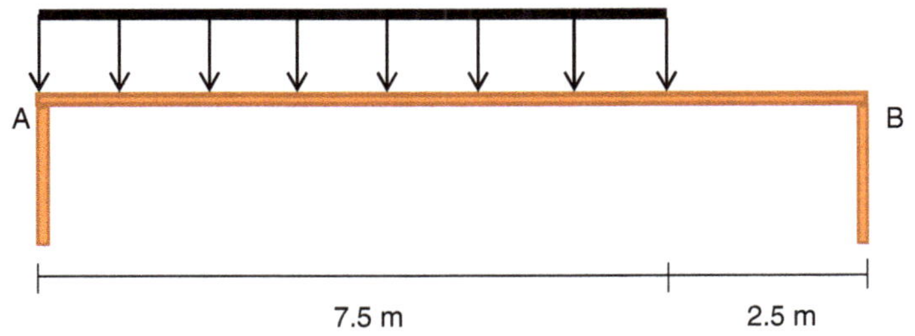

Solution:
$V_{y_{\max}} = 5.625\,\text{kN}$, the beam will not break

Goal:
Find the maximum internal force along the beam

Given:
Distributed load and geometry

Assumptions:
Distributed load

Equations Needed:

$$\sum F_x = 0, \sum F_y = 0, \sum M_z = 0$$

Answer:
FBD beam

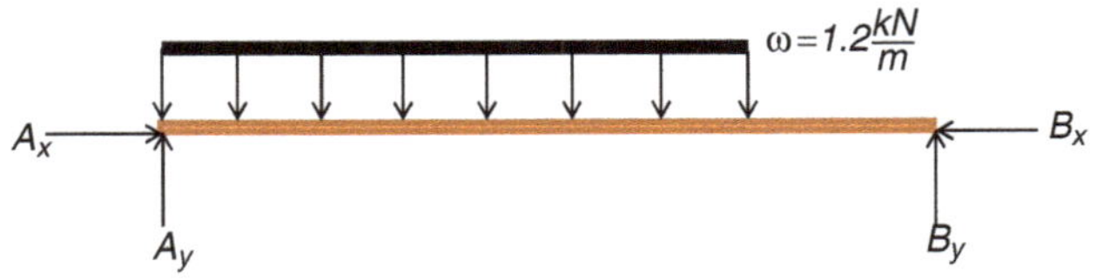

1) $\sum F_y = 0 = A_y - \omega \times 7.5\,\text{m} + B_y$

2) $\sum M_{z@A} = 0 = B_y \times 10\,\text{m} - \omega \times 7.5\,\text{m} \times \frac{7.5}{2}\,\text{m} \rightarrow B_y = 2.8125 \times \omega = 3.375\,\text{kN}$

From equation (1), $A_y = 4.6875 \times \omega = 5.625\,\text{kN}$

For $0 < x < 7.5$,

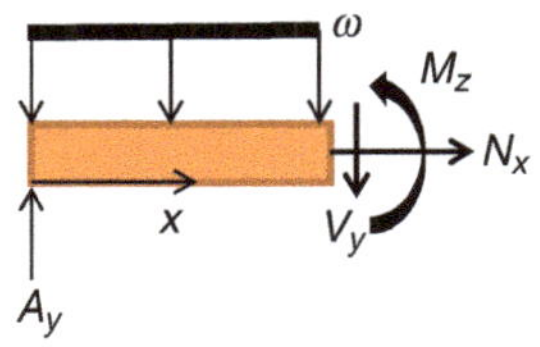

$$\sum F_y = 0 = A_y - \omega \times x - V_y \rightarrow V_y = \omega(4.6875 - x) = 5.625 - 1.2x\,[\text{kN}]$$

For $7.5 < x < 10$,

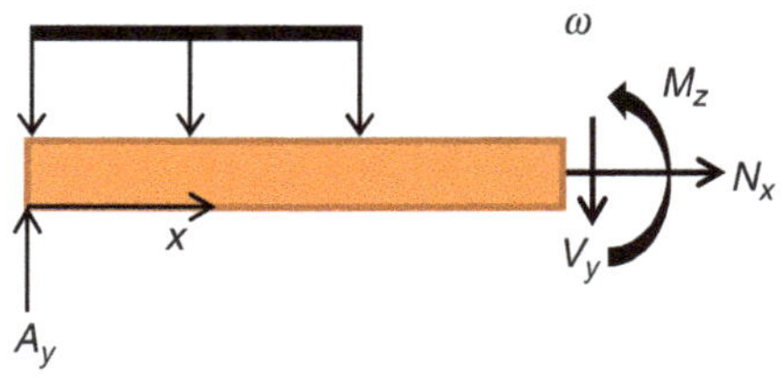

$$\sum F_y = 0 = A_y - \omega \times 7.5\,\text{m} - V_y \rightarrow V_y = -\omega(2.8125) = 3.375\,\text{kN}$$

Since the shear force is less than the maximum limit of 6 kN for both regions, the beam should not break if all attendees are stationary on the beam.

7.2. *V* AND *M* DIAGRAMS OF A COMPLEXLY LOADED BEAM

Problem Author: Yangling Zhou

Type of Problem:
Shear force and bending moment diagrams.
Know how to replace the force by a force and couple system and to draw shear force and bending moment diagram.

Problem Statement:
Draw Shear force and bending moment diagrams of the beam.

Drawing:

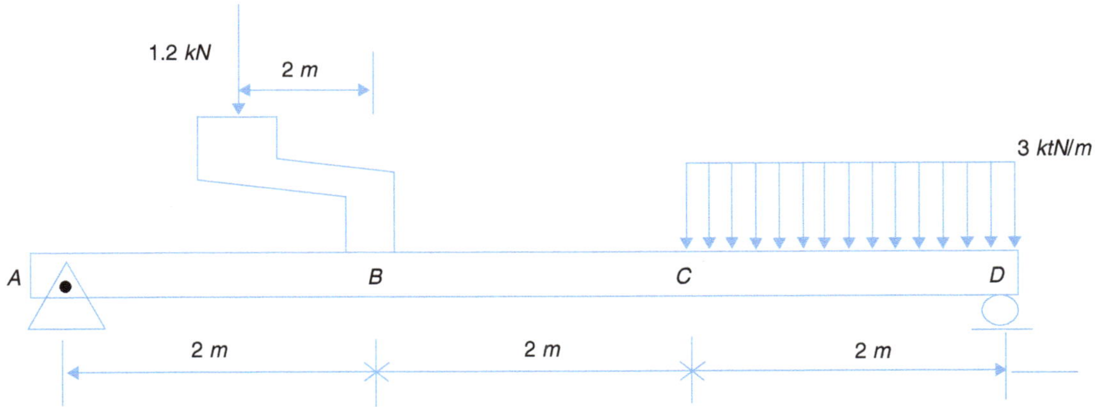

Goal:
Drew shear force and bending moments

Given:
External forces

Assumptions:
Beam is massless

Equations Needed:

$$\sum F_x = 0,$$

Answer:

FBD:

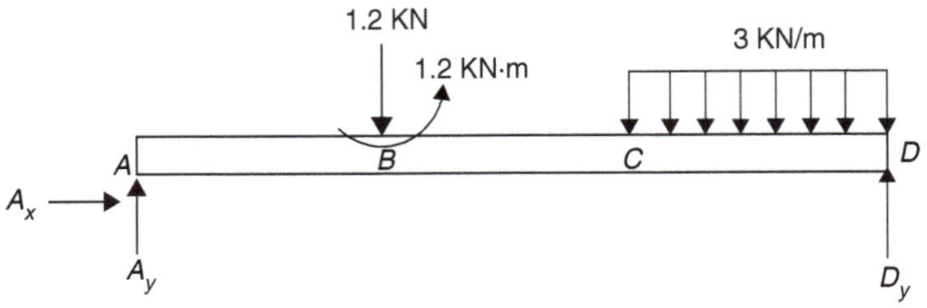

$$\sum F_x = A_x = 0 \Rightarrow A_x = 0$$

$$\sum M_A = -1.2 \cdot 2 + 1.2 - (3 \cdot 2) \cdot 5 + D_y \cdot 6 = 0 \Rightarrow D_y = 5.2 \text{ kN}$$

$$\sum F_y = A_y - 1.2 - (3 \cdot 2) + D_y = 0 \Rightarrow A_y = 2.0 \text{ kN}$$

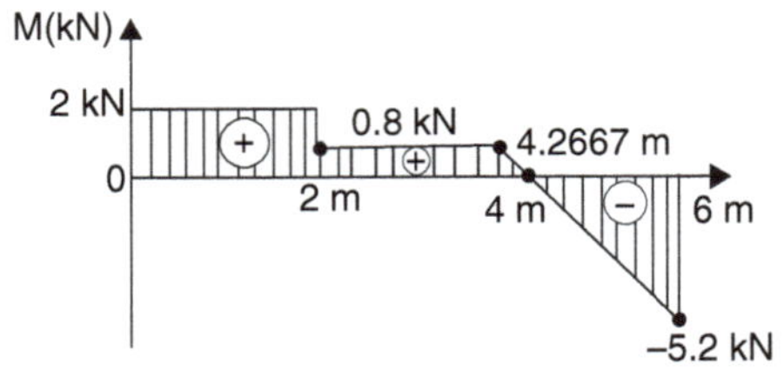

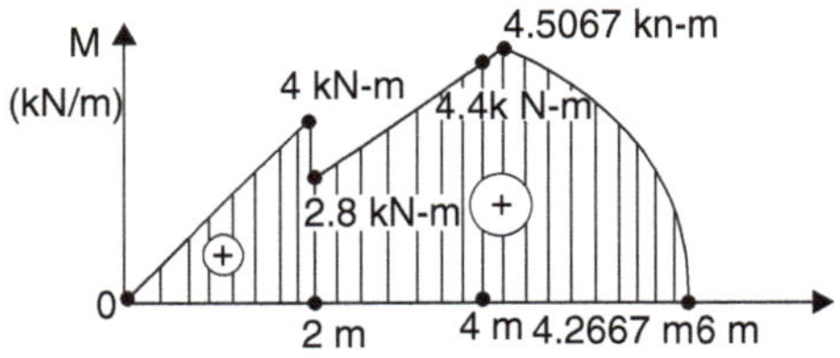

7.3. *V* AND *M* DIAGRAMS OF A BEAM #2

Problem Author: Keith R. Coleman

Type of Problem:
Shearing/Bending-Moment Diagrams

Problem Statement:
Develop the shear and bending moment diagrams from the beam in static equilibrium shown.

Assumptions:
Neglect the mass of the beam

Equations Needed:
Equations of Equilibrium

$$\Sigma \boldsymbol{F}_y = 0 \quad \Sigma \boldsymbol{F}_x = 0 \quad \Sigma M_o = 0$$

Moment

$$M = \boldsymbol{F}_\perp \cdot d$$

Where d is some distance away from an axis of rotation and where $\boldsymbol{F}_\perp$ is a perpendicular force that forms a 90° angle with the distance

Slope/Point-Slope Formula

$$m = \frac{y_2 - y_1}{x_2 - x_1}$$

$$y - y_1 = m(x - x_1)$$

Area (90° triangle)

$$A = \frac{1}{2} \cdot b \cdot h$$

Area (rectangle)

$$A = b \cdot h$$

Drawing:

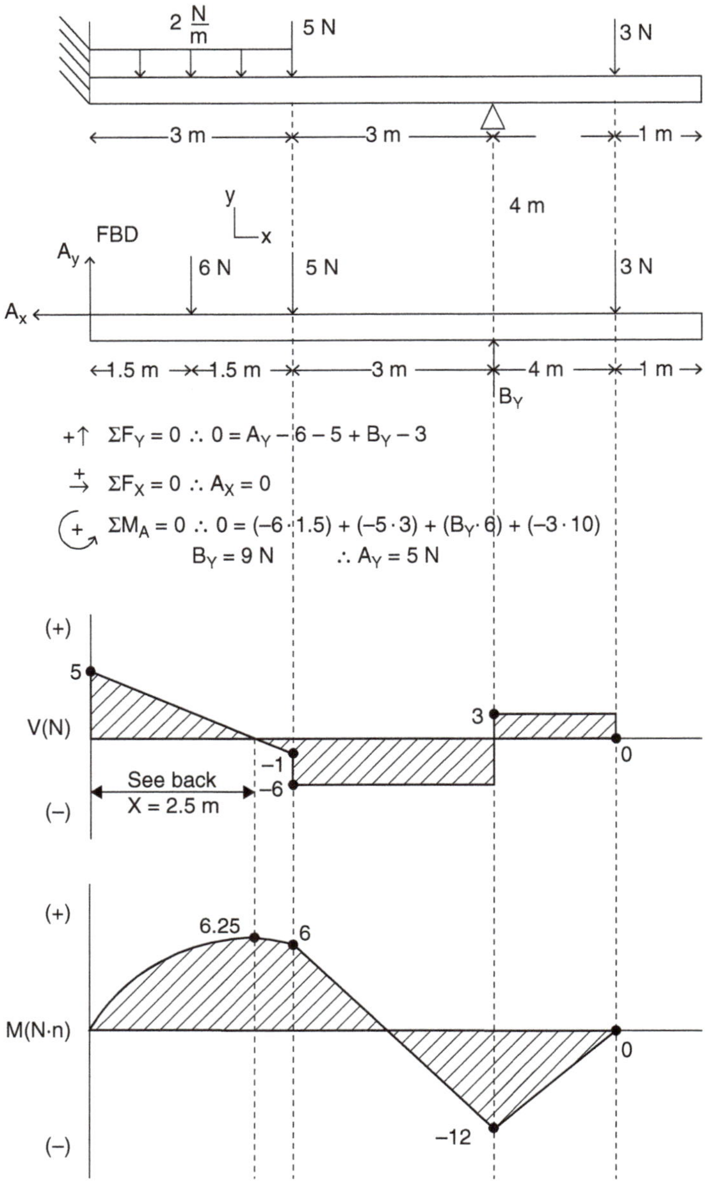

Slope Formula
Point-slope Formula

$$(0_m, 5_N) \quad (3_m - 1_N)$$

$$m = \frac{Y_2 - Y_1}{X_2 - X_1} = \frac{-1-5}{3-0} = \frac{-6}{3} = -2$$

$$Y - Y_1 = m(x - x_1)$$
$$Y - 5 = -2(x - 0)$$
$$Y - 5 = -2x$$
$$y = -2x + 5$$

When

$$y = 0$$

$$0 = -2x + 5$$

$$\therefore x = 5/2 \text{ or } 2.5 \text{ m}$$

Moment

$$0-2.5\text{ m: } A_{\Delta} = \frac{1}{2}\text{bh} = \frac{1}{2}(2.5\text{ m})(5\text{ N}) = 6.25\text{ N}\cdot\text{m}$$

$$2.5-3\text{ m: } A_{\Delta} = \frac{1}{2}\text{bh} = \frac{1}{2}(3-2.5\text{ m})(-1\text{ N}) = -0.\,25\text{ N}$$

$$3-6\text{ m: } A_{\square} = \text{bh} = (3\text{ m})(-6\text{ N}) = -18\text{ N}$$

$$6-10\text{ m: } A_{\square} = \text{bh} = (4\text{ m})(3\text{ N}) = 12\text{ N}$$

Solution:

See the two graphs V(N) and M(N·m) in the preceding image for final solution.

Explanation:

1) From the given scenario, we first develop an FBD.
2) Because the left-hand side is fixed and I randomly designated this point "*A*", there'll be two reaction forces, A_x and A_y. The pivot 6 meters away, which I designated as point "*B*," has only one reaction force, B_y.
3) With 3 unknowns, we employ three equations of static equilibrium.
4) Solving for those 3 unknown reaction forces now allows us to use the originally drawn scenario while imposing the 3 reaction forces onto in. Drawing lightly dashed lines vertically down the page from each segment where a new force is introduced along the beam enables us to more clearly develop the *M–V* diagrams.
5) Shear–Moment Diagram: There are two things to note primarily when developing these graphs by hand; how does the curve look and by how much does it increase/decrease. To answer the first question, see the following relationship:

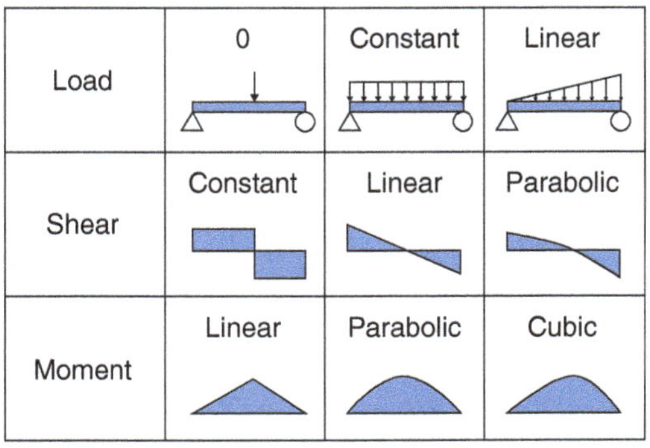

Essentially, at the "Load" level as seen in the originally given problem drawing, integration once gives the curve characteristic of shear. Integration once more gives the curve depiction of the moment.

6) Because A_y = 5 N, Shear diagram starts at +5 N, but because the next load is a constant uniformly distributed load, we know the shear will appear linearly. We also know because the distributed load is negative in magnitude, the linear line will have a negative slope. But by how much does it decrease?
7) Since the load is resembles a rectangle, we multiply base and height to get the magnitude of a representative resultant force, –6 N. This is how the linear curve decrease from +5 N to –1 N.
8) For the remaining point-load forces, shearing simply increases or decreases vertically accordingly. The segments between these point-load forces stay horizontally constant, as there are no forces changing their shearing magnitude.
9) Once shear diagram is complete, the moment diagram is completed the same way (see image 2 for moment calculations).
10) *Note:* Because the direction of the shearing force transitions from positive to negative in the region 0–3 m, the moment diagram must too follow, peaking at the length beam where the shear curve = 0 N. But what is the value of the length? There are two ways to find this. One is the point-slope method as shown in image 2. The other way, and probably more widely accepted method is simply use integration.

7.4. *V* AND *M* DIAGRAMS OF A BEAM #3

Problem Author: Derek S. Bibeault

Type of Problem:

The student should be able to bring their knowledge from previous sections in order to solve for internal reactions and construct a shear and moment diagram accordingly.

Problem Statement:

A horizontal steel beam *AB* is subjected the external loadings shown. Determine (a) the support reactions at *A* and *B* and (b) construct both a shear and moment diagram for the beam using the information found. *Hint:* Slug is not units of force!

Drawing:

Beam AB

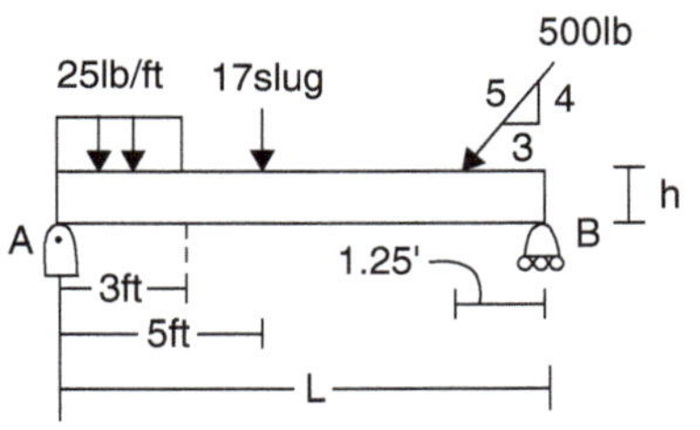

Solution:

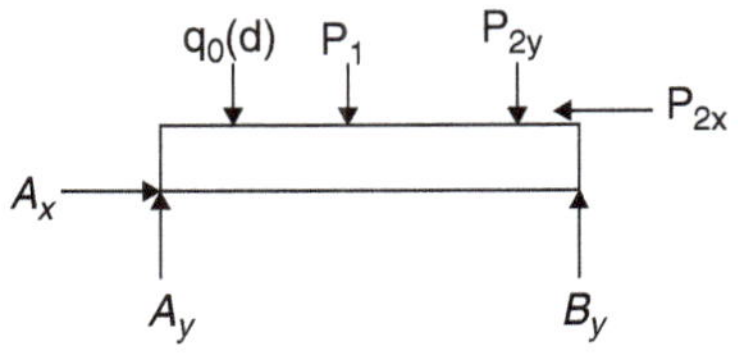

Equilibrium Equations

$$\sum F_x = 0$$
$$\sum F_y = 0$$
$$\sum M_x = 0$$

$$q_0 = 25 \text{ lb/ft}$$
$$\text{P}_1 = 17 \text{ slug(g)}$$
$$\text{P}_2 = 500 \text{ lb}$$

1) Draw FBD 8 write equations
2) Solve for unknowns

$$\sum F_x = A_x - (500)\left(\frac{3}{5}\right) = 0$$

$$A_x = 300 \text{ lb}$$

$$\sum F_y = A_y + B_y - \left(25\ \frac{\text{lb}}{\text{ft}}\right)(3\text{ ft}) - (17\text{ slug})\left(32.2\frac{\text{ft}}{5^2}\right) - 500\text{ lb}\left(\frac{4}{5}\right) = 0$$

$$A_y + B_y = 75\text{ lb} + 547.4\text{ lb} + 400\text{ lb}$$

$$A_y + B_y = 1022.4$$

$$\sum M_B = -11\text{ ft}(A_y) + 75\text{ lb}(9.5\text{ ft}) + 547.9\text{ lb}(6\text{ ft}) + 400\text{ lb}(1.25\text{ ft}) + 300\text{ lb}(1\text{ ft}) = 0$$

$$A_y = \frac{-712.5\text{ lb}\cdot\text{ft} - 3284.4\text{ lb}\cdot\text{ft} - 500\text{ lb}\cdot\text{ft} - 300\text{ lb}\cdot\text{ft}}{-11}$$

$$A_y = 436.1\text{ lb}$$

$$B_y = 586.3\text{ lb}$$

*751b is the distributed load acting at its centrold

3) Show diagram of Internal and external reactions

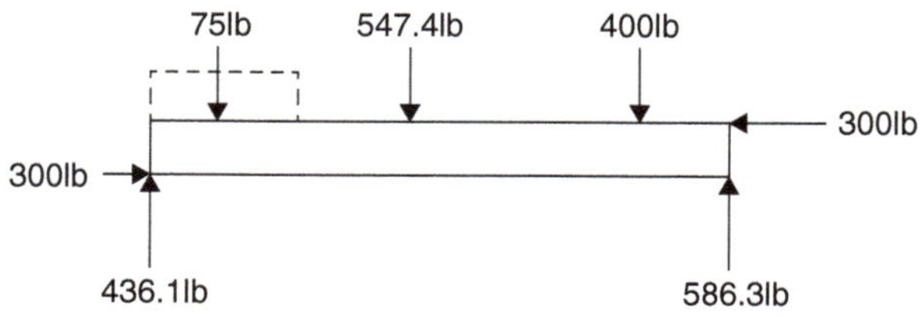

4) Draw shear & Moment Diagrams

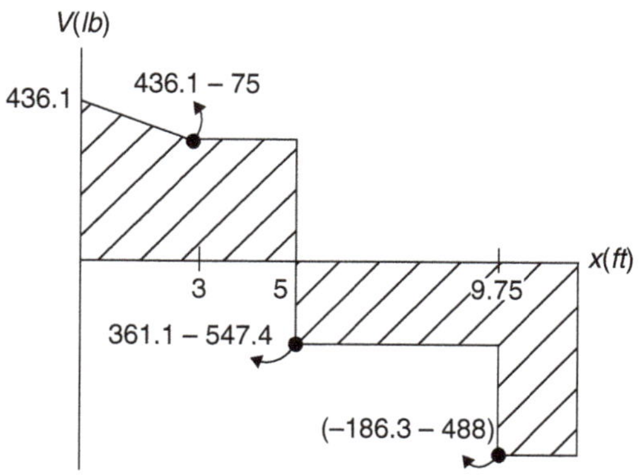

Notes: *When shear increases or decreases linearly, the bending moment increases or decreases exponentially

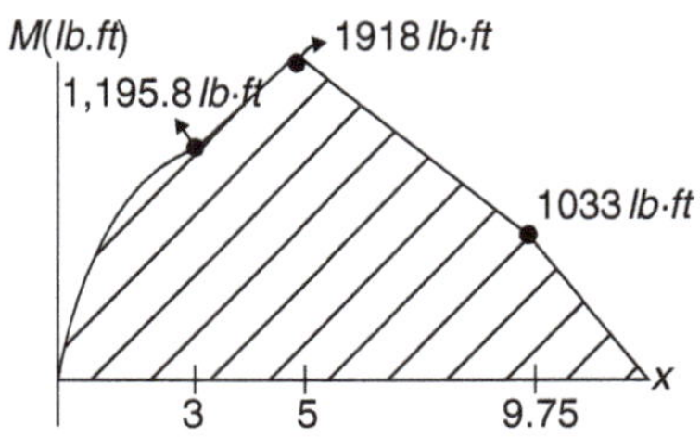

*The moment at each point can be determined by calculating the area of the shear graph. Or by basic beam analysis, as shown below.

Moment @ 3ft

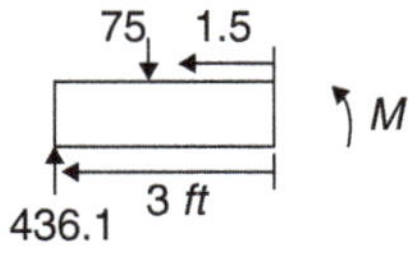

$$\Sigma M = M - 436.1(3) + 75(1.5) = 0$$
$$M = 1308.3 - 112.5$$
$$M@3^1 = 1{,}195.8 \text{ lb}\cdot\text{ft}$$

Moment @ 5ft

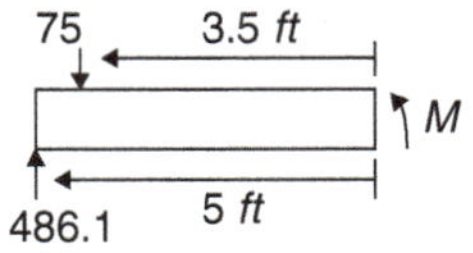

$$\Sigma M = M - 436.1(5) + 75(3.5) = 0$$
$$M = 2180.5 - 262.5$$
$$M@5 \text{ ft} = 1918 \text{ lb}\cdot\text{ft}$$

Moment @ 9.75ft

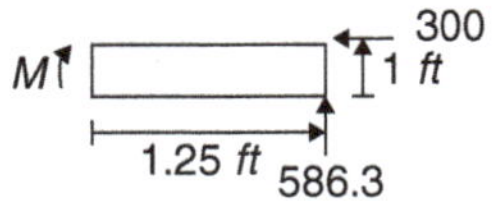

$$\Sigma M = -M + 300(1) + 586.3(1.25) = 0$$
$$M = 300 + 732.8$$
$$M@9.75 \text{ ft} = 1033 \text{ lb}\cdot\text{ft}$$

Goal:

Pull together your knowledge of statics in order to answer this difficult problems which includes multiple parts and concepts.

Given:

L = 11 ft., h = 1 ft., pin connected at A, roller at B

Assumptions:

The weight of the beam itself is neglected. The beam is in static equilibrium (obviously).

Equations Needed:

$$\sum \boldsymbol{F}_x = 0$$

$$\sum \boldsymbol{F}_y = 0$$

$$\sum \boldsymbol{M}_i = 0$$

7.5. BEAM ANALYSIS WITH A CIRCUS ELEPHANT

Problem Author: Prof. Jamie Gurganus

Problem Statement:

At a circus act, a 6,000 lb. (26,689 N) elephant is standing on a 7.3 m beam being held by ball (roller) and a pin. Find the reaction forces and plot the shear and bending moment diagram. Assume the elephant feet each hold half the weight.

You will need to solve for the distributed load in each pair of feet. Assume each of the elephant feet hold half the weight.

Drawing:

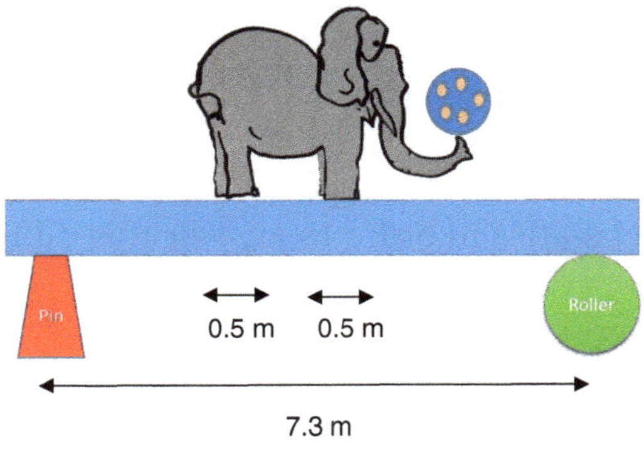

Solution:

Step I: Draw the FBD. The distributed load of the elephant is given:

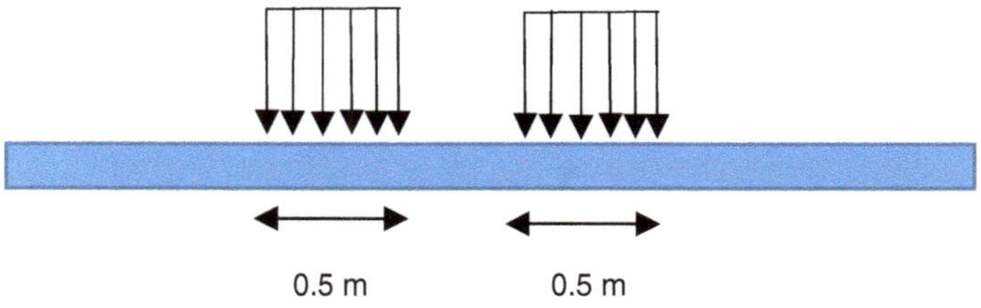

Step II: Use the equations of equilibrium to solve for the reaction forces.

Step III: Solve and plot the shear and bending moment diagram.

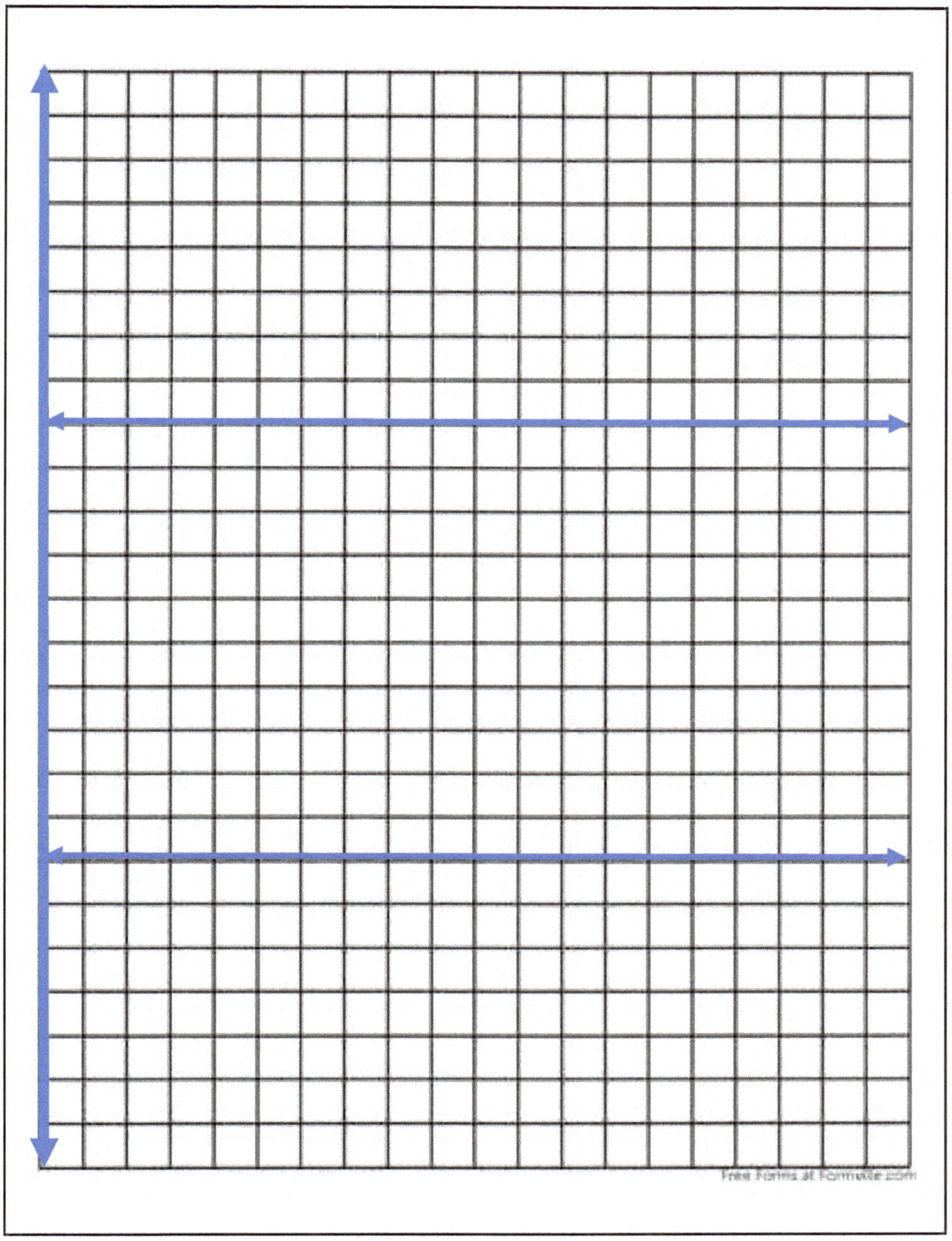

CHAPTER 8

Moment of Inertia of an Area

DEFINITION OF MOMENT OF INERTIA (MOI) OF AN AREA

Moments of inertia (MOI) are not necessarily used in statics. You will likely see this concept in your mechanics of materials class. However, calculations for moments of inertia are similar to finding centroids. Moments of inertia are used in courses such as mechanics of materials, dynamics, and fluid mechanics.

The *MOI* of an object is a measure of its resistance to change in rotation.

Think about this . . .

Recall driving your car or watching someone drive. Experience tells us that it is harder to start (or stop) a large wheel turning than a small wheel. Mathematically, this is represented by the large wheel having a larger MOI.

Another example is shown below.

Recall: Area MOI: a measure of how easy or hard it is to bend something.

Beam with load (less surface area)

Same beam with load on its flat side (more surface area)

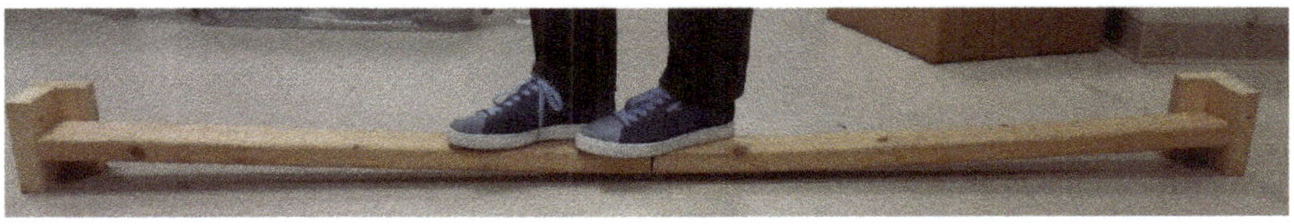

Notice the beam bends or deflects easier on one side versus another. Think about it. Why? What has changed? Which beam has the higher MOI?

MOI are used in various engineering calculations, including:

- Calculating stresses in beams—they are at times related to the MOI of the cross-sectional area of the beam (resistance to bending) (mechanics of materials)
- Mass moments of inertia are used in studying the rotational motion of objects (dynamics)

CALCULATING AREA MOMENTS OF INERTIA

Recall from calculus that Integrals are in the form shown below. These are known as the first moments of the area.

$$\int x\,dA$$

In studying the second moments of the area or area moments of inertia, you will see the integral in this form:

$$\int x^2 dA$$

Moment of inertia about *x*-axis	$I_y = \int_A x^2 dA$
Moment of inertia about *y*-axis	$I_x = \int_A y^2 dA$
Polar moment of inertia (*z*-axis)	$J_o = \int_A r^2 dA$

RADIUS OF GYRATION

The MOI is sometimes expressed in terms of the radius of gyration. The radius of gyration determines how the area is distributed around the centroid.

$$I_x = k_x^2 A \, I_y = k_y^2 A$$

Parallel Axis Theorem: If you know the MOI about a centroidal axis of a figure, you can calculate the MOI about any *parallel* axis to the centroidal axis using the following formulas:

$$I_x = I_{x'} + d_y^2 A$$

$$I_y = I_{y'} + d_x^2 A$$

$$J_o = J_c + d_x^2 A$$

where $I_{x'}$, $I_{y'}$, J_c = initial centroidal value (generally found in an Appendix of a statics book. A few shapes are included in this chapter).

STEPS TO SOLVING MOMENTS OF INERTIA

Integration Method

For this method you are welcome to use double integration for an element. However, for simplicity, the area of the element with a differential size in only one direction can be used (dx or dy).

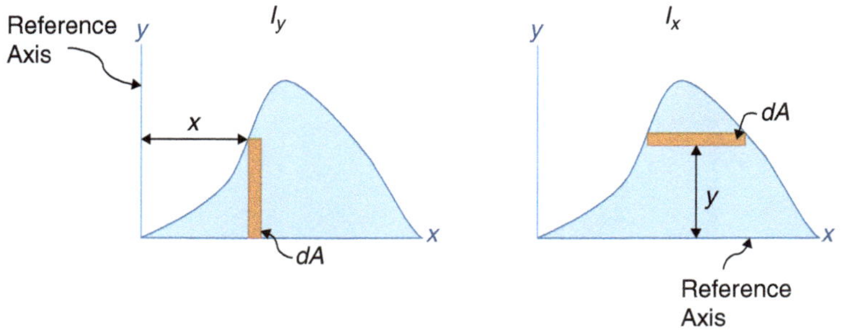

1. Choose the element dA. There are two choices: a vertical strip or a horizontal strip.
2. Integrate to find the MOI.

Composites

An I-Beam, designed to have a high MOI, is a common composite shape you will see in engineering applications. To determine the MOI it is best to break it down into simpler shapes. Two examples of calculating the MOI of an I-beam are shown in the following.

Both examples, the MOI is determined calculating the addition (or subtraction) of the various rectangles. In the first example, rectangles 1, 2, and 3, MOI is added using the known formula for the MOI of a rectangle and the parallel axis theorem, since 2 and 3 lie away from the central axis.

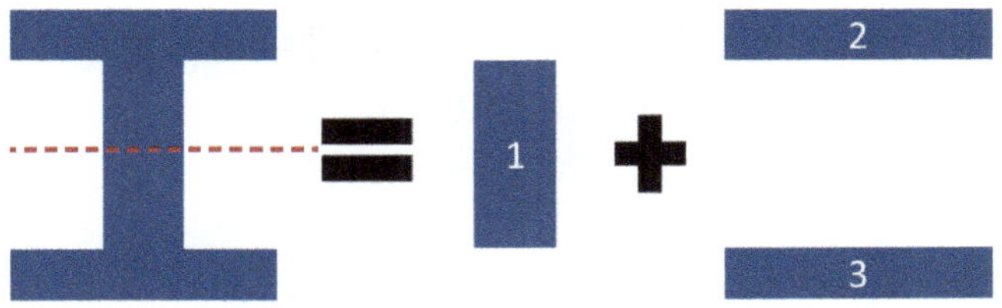

The second example, the MOI of each of the rectangles, 2 and 3, are subtracted from rectangle 1. This time, however, the parallel axis theorem is not needed.

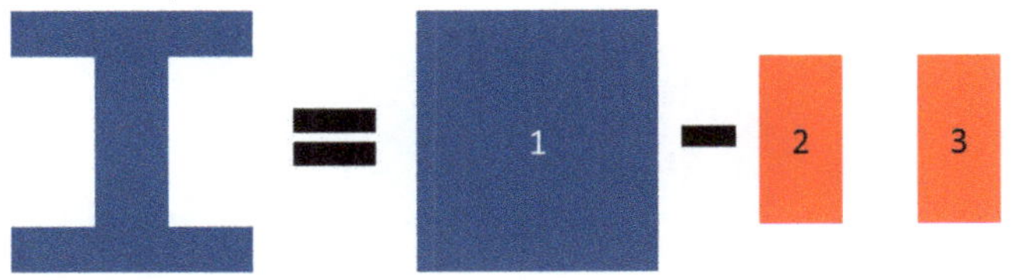

If you have a simple shaped object, many textbooks have the MOI solved. This makes solving for a composite shape easier. Common shapes used are shown on the following page.

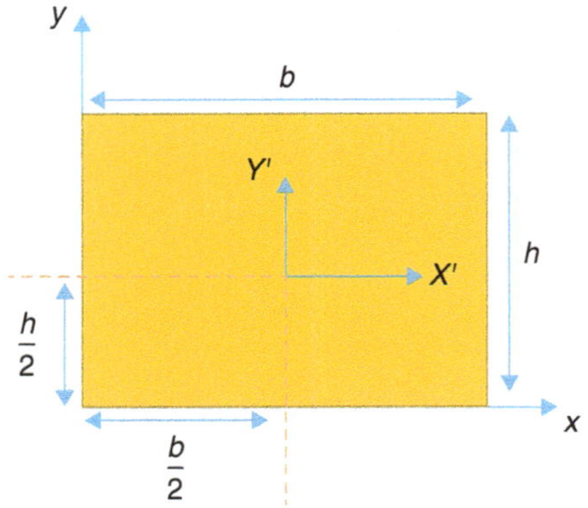

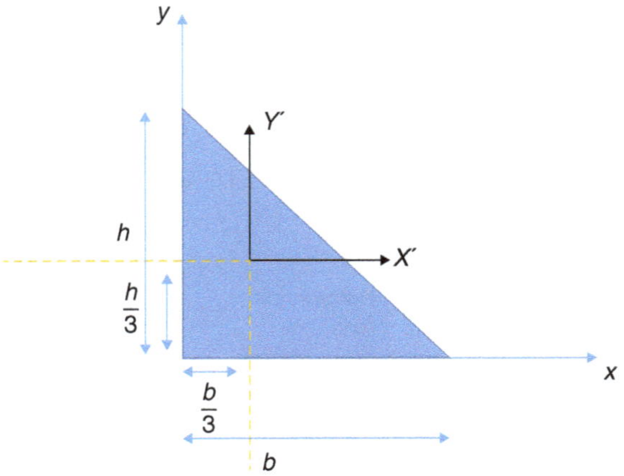

$$I_y = \frac{1}{12}b^3h$$

$$I_x = \frac{1}{12}bh^3$$

Steps to solving MOI of composites (using presolved MOI):

1. Divide the object given area into its simpler shaped parts.
2. Locate the centroid of each part and indicate the perpendicular distance from each centroid to the desired reference axis.
3. If you wish to find the MOI about the centroidal axes of the entire object, find the centroid of the entire object. If you wish to find the MOI about fixed axes, proceed to the next step.
4. Determine the MOI of each simpler shaped part (generally about its centroid) using the table provided in a text book.
5. Use the parallel axis theorem to find the desired axes.
6. The MOI of the entire object is the sum of the MOI's for each simpler shape. Don't forget that the MOI is negative for shapes to be subtracted.

SECTION 8 Problems

8.1. MOMENT OF INERTIA OF A HOCKEY STICK

Problem Author: Richard Blorstad

Problem Statement:
In ice hockey, some players, especially forwards who like to shoot a lot, prefer sticks that are more flexible, while others, like defensemen, may prefer stiffer sticks, though not too stiff. You may assume the cross section is rectangular and that the width (w) is twice the thickness (t). What is the thickness and width required for the minimum MOI to equal 0.407 in.^{4}?

Drawing:

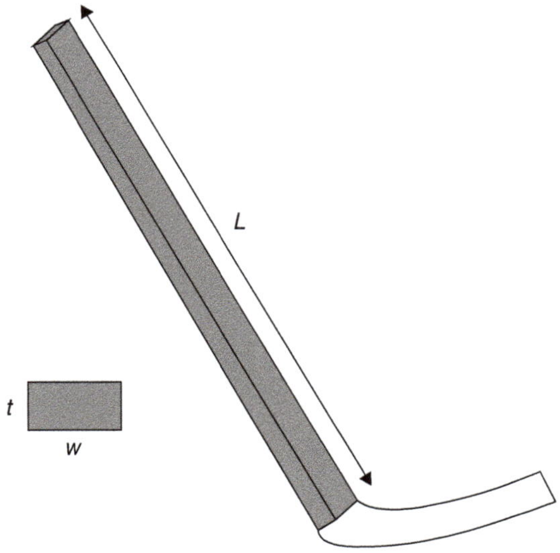

Solution:

Goal:
Solve for the thickness and width of the hockey stick cross section, given the MOI.

Given:

MOI = 0.407 in.^{4} and the cross section shown, where $2t = w$.

Assumptions:

None

Equations Needed:

$$I = \frac{1}{12} bh^3$$

Answer:

For the central axis that is parallel to the width

$$I = \frac{1}{12} wt^3 = 0.407 \text{ in.}^4$$

$$\frac{1}{12}(2t)(t)^3 = 0.407 \text{ in.}^4$$

$$\frac{1}{6}t^4 = 0.407 \text{ in.}^4$$

$$t = (0.407 \times 6 \ \text{ in.}^4)^{1/4}$$

$$t = 1.25 \text{ in.}$$

$$w = 2t = 2.50 \text{ in.}$$

For the central axis parallel with the thickness

$$I = \frac{1}{12} wt^3 = 0.407 \text{ in.}^4$$

$$\frac{1}{12}(t)(2t)^3 = 0.407 \text{ in.}^4$$

$$\frac{2}{3}t^4 = 0.407 \text{ in.}^4$$

$$t = \left(0.407 \times \frac{3}{2} \text{in.}^4\right)^{1/4}$$

$$t = 0.884 \text{ in.}$$

$$w = 2t = 1.77 \text{ in.}$$

If the thickness is 0.884 in., then the MOI along the central axis parallel to the width is too low. Therefore the required thickness and width are 1.25 in. and 2.50 in., respectively.

8.2. PARALLEL AXIS THEOREM

Problem Author: Caroline Zellhofer

Type of Problem:
This is a parallel axis theorem problem, which challenges students to find the second moment of area about the x- and y-axis using standard derived area moments of inertia.

Problem Statement:
An L-beam is constructed from two 10 cm by 40 cm wooden boards. Using the given coordinate axis, find the second moment of area, I, about the centroid in both the x and y directions using the parallel axis theorem.

Drawing:

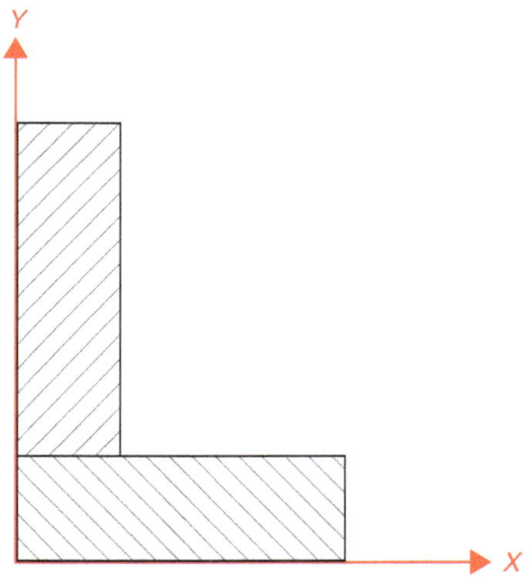

Solution:

1) Find the centroid of the cross-section.
   a) Find the distance from the x-axis to the x'–x' axis, $\bar{x}$.
      i) Break up the cross-section into numbered parts, for organization.

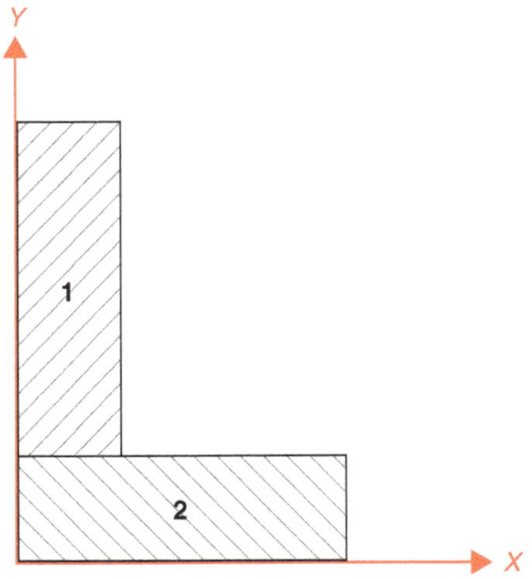

ii) Follow the equation $\bar{x} = \frac{\sum x_i A_i}{\sum A_i}$, where x_i is the distance from the x-axis to the centroid of the part, and A_i is the area of the part. (Hint: using a table to organize your data may help keep computations straight).

1) *Note*: all distances are in cm.

$$x_1 = 10\,\text{cm} + 20\,\text{cm} = 30\,\text{cm}$$

$$x_2 = 5\,\text{cm}$$

Part Number	A_i	$x_i A_i$	$\bar{x} = \frac{\sum x_i A_i}{\sum A_i}$
1	(10)(40) = 400	(30)(400) = 12,000	
2	400	(5)(400) = 2,000	
Total	800 cm^2	14,000 cm^3	(14,000)/(800) = 17.5 cm

iii) Therefore $\bar{x} = 17.5$ cm. Thus, the x'–x' axis is 17.5 cm from the x-axis.

b) Find the distance from the y-axis to the y'–y' axis, $\bar{y}$.

i) Following the same process as before, but for the y-axis:

1) *Note*: all distances are in cm.

$$y_1 = 5\,\text{cm}$$

$$y_2 = 20\,\text{cm}$$

(*Note*: be sure to be careful of the change in geometry from the x-axis to the y-axis. Don't just assume the numbers will be the same!)

Part Number	$\sum A_i$	$\sum y_i A_i$	$\bar{y} = \frac{\sum y_i A_i}{\sum A_i}$
1	(10)(40) = 400	(5)(400) = 2,000	
2	400	(20)(400) = 8,000	
Total	800 cm^2	10,000 cm^3	(10,000)/(800) = 12.5 cm

ii) Therefore $\bar{y} = 12.5$ cm. Thus, the y'–y' axis is 12.5 cm from the y-axis.

c) The centroid on the diagram appears as follows:

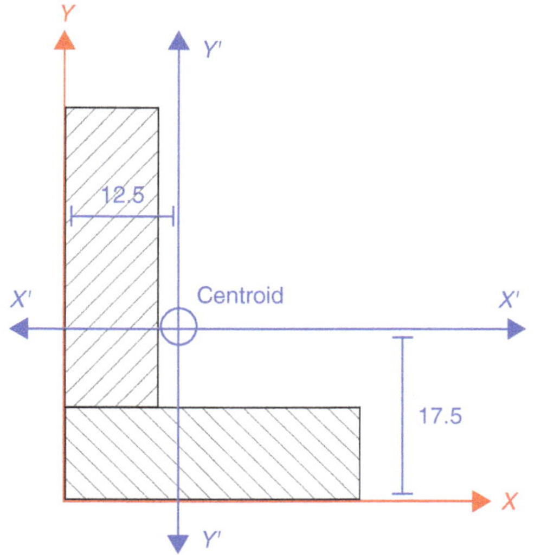

2) Find I_{xx} using parallel axis theorem.

a) Apply $I_{xx} = \sum (I_x + A_i d_i^2)$, where I_x is the second moment of area given by the specific shape of the part, A_i is the area of the part, and d_i is the distance from the centroid of the entire cross-section to the centroid of the part with respect to the x- or y-axis.

Note: be especially careful of calculating d_i! It is an easy step to miscalculate. Again, a chart would be useful here to organize the calculations.

For a rectangle,

$$I_x = I_y = \frac{bh^3}{12}$$

Part Number	I_x	A_i	d_i	$I_x + A_i d_i^2$
1	$(10)(40^3)/12 =$ 53,333.3	400	(17.5 – 30) = –12.5	(53,333.3) + (400) $(12.5)^2$ = 115,833.3
2	$(40)(10^3)/12 =$ 3,333.3	400	(17.5 – 5) = 12.5	(3,333.3) + (400) $(12.5)^2$ = 65,833.3
Total				181,666.6 cm^4

b) Therefore $I_{xx} = 181{,}666.6\text{ cm}^4$.

3) Find I_{yy}, following the same steps as before.

a)

Part Number	I_y	A_i	d_i	$I_y + A_i d_i^2$
1	$(40)(10^3)/12 =$ 3,333.3	400	$(12.5 - 5) = 7.5$	$(53{,}333.3) + (400)(7.5)^2 = 75{,}833.3$
2	$(10)(40^3)/12 =$ 53,333.3	400	$(12.5 - 20) = 7.5$	$(3{,}333.3) + (400)(7.5)^2 = 25{,}833.3$
Total				101,666.6 cm^4

b) Therefore $I_{yy} = 101{,}666.6\text{ cm}^4$.

Goal:

To find I_{xx} and I_{yy} about the centroid.

Given:

Dimensions of L-beam

Assumptions:

None

Equations Needed:

$$\bar{x} = \frac{\sum y_i A_i}{\sum A_i}$$

$$I_{xx} = \sum (I_x + A_i d_i^2)$$

$$I_x = I_y = \frac{bh^3}{12}$$

Answer:

$$I_{xx} = 181{,}666.6\text{ cm}^4$$

$$I_{yy} = 101{,}666.6\text{ cm}^4$$

8.3. MOMENT OF INERTIA OF A COMPOSITE SHAPE

Problem Author: Prof. Jamie Gurganus

Type of Problem:

Find the MOI for the area about the y-axis.

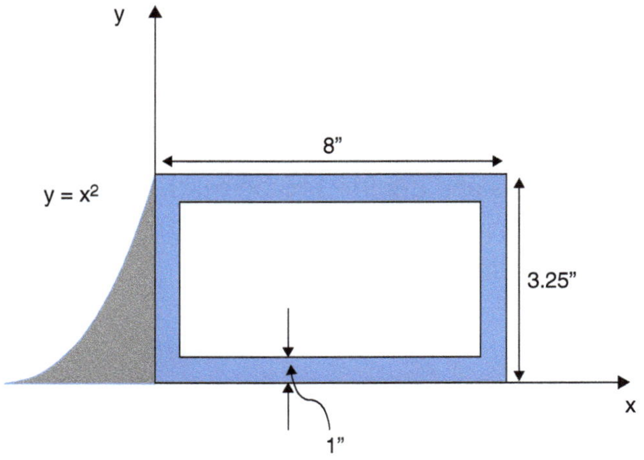

8.4. I-BEAM ANALYSIS

Problem Author: Caroline Zellhofer

Type of Problem:

This is a parallel axis theorem problem, which challenges students to find the second moment of area about the x- and y-axis using standard derived area moments of inertia.

Problem Statement:

For design specifications, an I-beam is required to have a second moment of area, I_{xx}, greater than 1.898e5 cm^4. The two flanges are decided to be made from two 5 cm by 30 cm bars. If the span of the web must be 40 cm, what is the required width of the web (labeled x in the diagram)? Then find I_{yy} using this information.

Drawing:

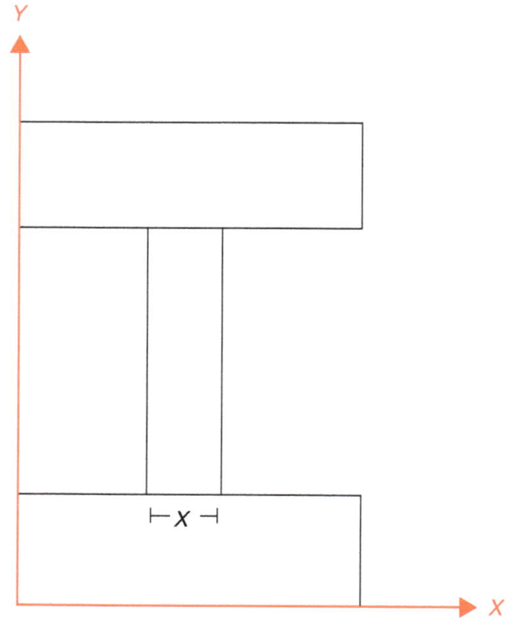

Solution

1) Find the centroid by inspection.
   a) Notice that the I-beam is symmetric in area in both the x- and y-direction.
   b) Thus, $\bar{x} = 5 + \frac{40}{2} = 25$ cm.
   c) And $\bar{y} = \frac{30}{2} = 15$ cm.

2) Using I_{xx}, find x using parallel axis theorem.

a) Apply $I_{xx} = \sum (I_x + A_i d_i^2)$, where I_x is the second moment of area given by the specific shape of the part, A_i is the area of the part, and d_i is the distance from the centroid of the entire cross-section to the centroid of the part with respect to the x- or y-axis.

Note: be especially careful of calculating d_i! It is an easy step to miscalculate. Again, a chart would be useful here to organize the calculations. Filling in the chart in terms of the variable x allows us to plug in I_{xx} later.

For a rectangle, $I_x = I_y = \dfrac{bh^3}{12}$

Part Number	I_x	A_i	d_i	$I_x + A_i d_i^2$
Web	$(x)(40^3)/12 =$ $5{,}333.3x$	$(x)(40) = 40x$	0	$5{,}333.3x$
Upper Flange	$(30)(5^3)/12 =$ 312.5	$(5)(30) = 150$	$(25 - 47.5) = -22.5$	$(312.5) + (150)(22.5)^2 = 76{,}250$
Lower Flange	312.5	150	$(25 - 2.5) = 22.5$	76,250
Total				$(152{,}500 + 5{,}333.3x)$ cm^4

b) Therefore $I_{xx} = 189{,}800 \text{ cm}^4 = 152{,}500 + 5{,}333.3x$

c) Solving for x, $x = 6.9937$ cm, or approximately 7 cm.

3) Find I_{yy}, following the same steps as before.

a)

Part Number	I_x	A_i	d_i	$I_x + A_i d_i^2$
Web	$(40)(7^3)/12 = 1{,}143.3$	$(7)(40) = 280$	0	1,143.3
Upper Flange	$(5)(30^3)/12 = 11{,}250$	$(5)(30) = 150$	0	11,250
Lower Flange	11,250	150	0	11,250
Total				23,643.3 cm^4

b) Therefore $I_{yy} = 23{,}643.3 \text{ cm}^4$.

Goal:

To find the width of the web, x, and I_{yy} about the centroid.

Given:

Dimensions of I-beam

I_{xx}

Assumptions:

Assume that the I-beam is symmetric in area in both the x- and y-direction.

Equations Needed:

$$I_{xx} = \sum (I_x + A_i d_i^2)$$

$$I_x = I_y = \frac{bh^3}{12}$$

Answer:

$$x = 7\,\text{cm}$$

$$I_{yy} = 23{,}643.3\ \text{cm}^4$$

8.5. MOMENT OF INERTIA OF A SHAPE WITH A HOLE

Problem Author: Prof.Jamie Gurganus

Type of Problem:

Find the MOI for the area about the x-axis.

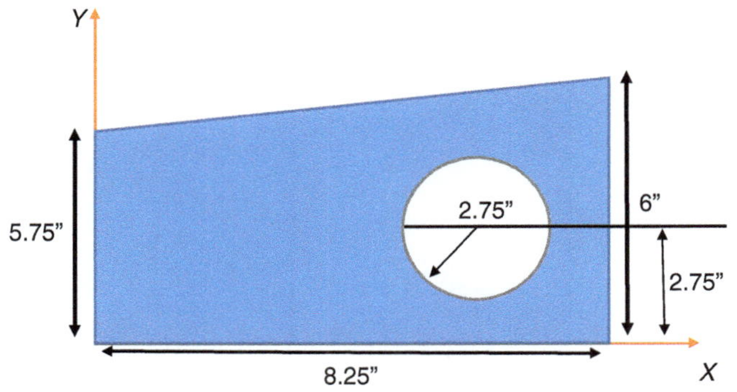

Solution:

Appendix

Tricks in Trig

SOHCAHTOA

$\sin\theta = \dfrac{\text{opposite}}{\text{hypotenuse}}$ $\cos\theta = \dfrac{\text{adjacent}}{\text{hypotenuse}}$

$\tan\theta = \dfrac{\text{opposite}}{\text{hypotenuse}}$

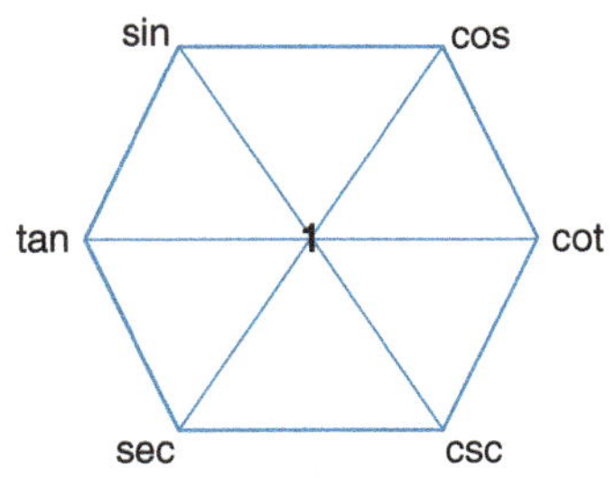

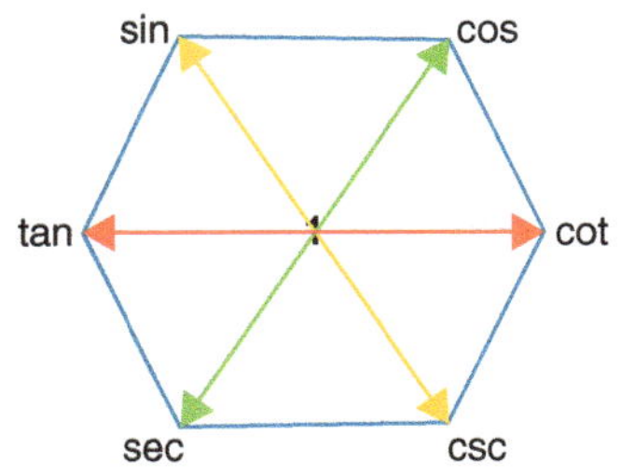

A trick to help remember your other identities

Use the diagram below (generally referred as the "magic hexagon").

There a few simple things to learn as this will help you remember every trig identity. Let's start with **sin***θ*. As you may recall from trigonometry $\sin\theta = 1/\csc\theta$. Notice on the hexagon, you simply start at **sin***θ* travel across the hexagon passing the "1" leaving you with $1/\csc\theta$.

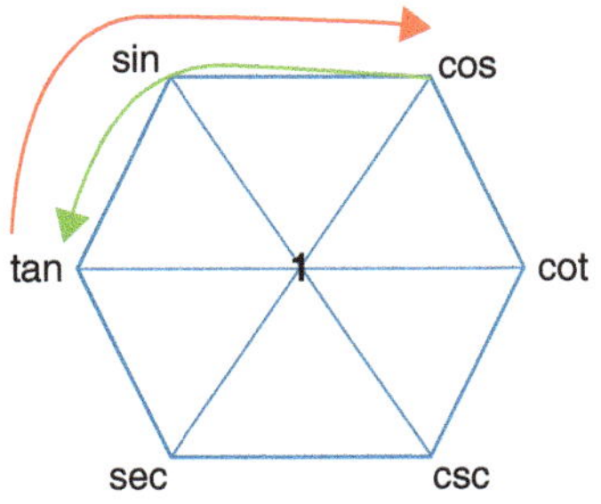

The next set of identities can be determined by going around the hexagon's outer edges. By traveling clockwise, $\tan\theta = \sin\theta/\cos\theta$, $\sin\theta = \cos\theta/\cot\theta$ etc. The same rule works if you travel counterclockwise, where, for example, $\cos\theta = \sin\theta/\tan\theta$. Additionally, any function on the outer edges of the hexagon is the products of the two adjacent functions, such as $\sin\theta = \cos\theta\tan\theta$, $\tan\theta = \sin\theta\sec\theta$, and the same applies around the entire hexagon.

If that hasn't yet amazed you, there are still two more sets of trig identities we can derive from the hexagon. First, by going across the hexagon horizontally, we can identify which trig functions are the same with complimentary angles. For example, $\sin\theta = \cos(90° - \theta)$, $\tan\theta = \cot(90° - \theta)$, and the same is true for secant and cosecant. Finally, by examining the three triangles that have a horizontal top edge, the Pythagorean Theorem holds true. By working clockwise around these three triangles, it can be shown that $\sin^2\theta + \cos^2\theta = 1$, $\tan^2\theta + 1^2 = \sec^2\theta$, and that $1^2 + \cot^2\theta = \csc\theta$.

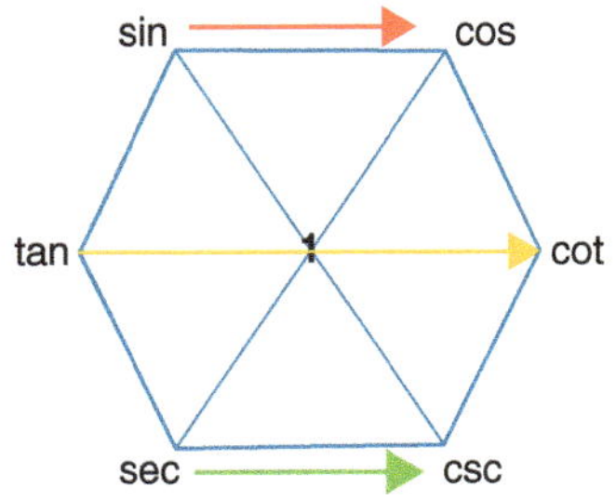

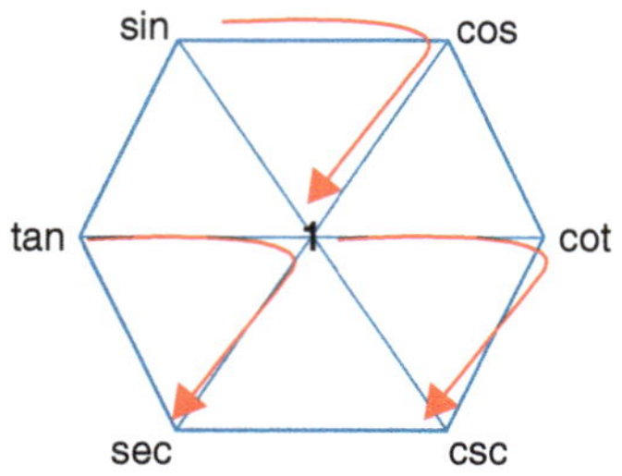

DETERMINANTS

For a determinant, one way to solve it is to extend the matrix. This allows you to add up all the diagonals from the top left to the bottom right (green) and subtract all the diagonals from the top left to the bottom left (red). This provides an answer for the determinant. For an n by n matrix, you have to add n–1 columns to the matrix.

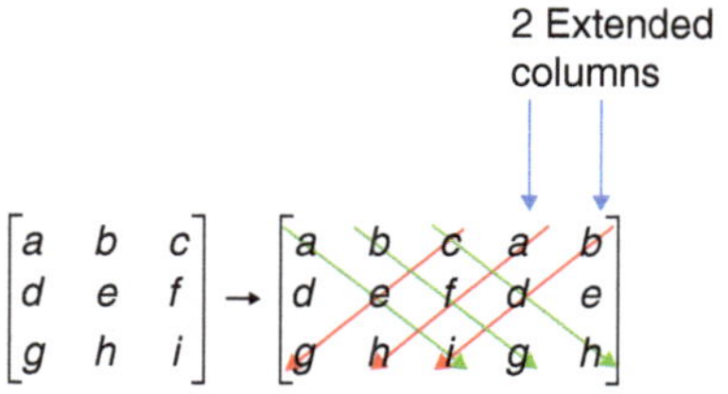

This is a 3 × 3 matrix. Therefore, we extend 3 – 1 = 2 columns. (n – 1)

(aei) + (bfg) + (cdh) – (ceg) – (afh) – (bdi)

www.ingramcontent.com/pod-product-compliance
Ingram Content Group UK Ltd.
Pitfield, Milton Keynes, MK11 3LW, UK
UKHW050722170926
15229UKWH00010B/48

9 781465 294401